George Washington Warren

A Compend of Dental Pathology and Dental Medicine

Containing the Most Noteworthy Points upon the Subjects of Interest to the Dental

Student and a Section on Emergencies. Second Edition

George Washington Warren

A Compend of Dental Pathology and Dental Medicine
Containing the Most Noteworthy Points upon the Subjects of Interest to the Dental Student and a Section on Emergencies. Second Edition

ISBN/EAN: 9783337255114

Printed in Europe, USA, Canada, Australia, Japan

Cover: Foto ©berggeist007 / pixelio.de

More available books at **www.hansebooks.com**

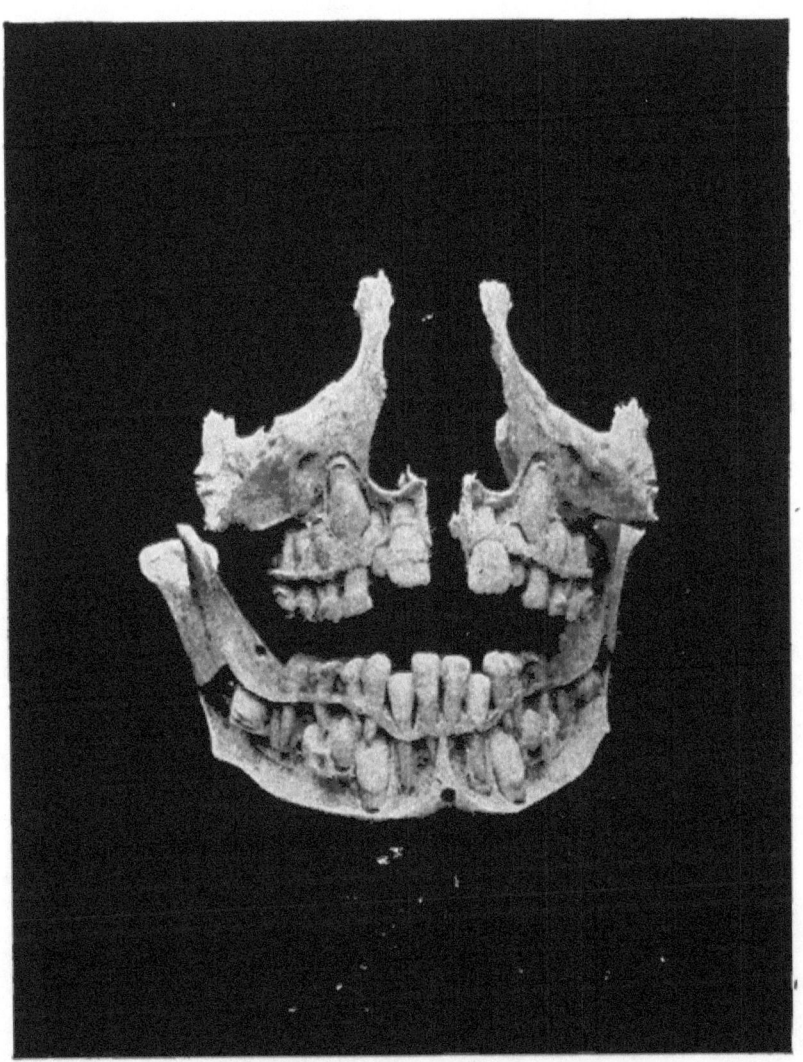

FRONTISPIECE.—A STUDY.

From a photograph of an excellent specimen, presented to the author by Dr. Amos H. Sibley.

?QUIZ COMPENDS? No. 13.

A COMPEND

OF

DENTAL PATHOLOGY

AND

DENTAL MEDICINE,

CONTAINING THE MOST NOTEWORTHY POINTS UPON THE SUBJECTS OF INTEREST TO THE DENTAL STUDENT AND A SECTION ON EMERGENCIES.

BY

GEO. W. WARREN, D.D.S.,

CHIEF OF THE CLINICAL STAFF, PENNSYLVANIA COLLEGE OF DENTAL SURGERY, PHILADELPHIA.

SECOND EDITION. ILLUSTRATED.

PHILADELPHIA:
P. BLAKISTON, SON & CO.,
No. 1012 WALNUT STREET.
1893.

Press of Wm. F. Fell & Co.,
1220-24 Sansom St.
Philadelphia.

DEDICATED TO

PROFESSOR C. N. PEIRCE, D. D. S.,

WHO, AS A TEACHER FOR OVER A QUARTER OF A CENTURY, HAS WITNESSED MUCH AGITATION AND MANY MODIFICATIONS IN PRACTICE, RESULTING IN THE PRESENT ADVANCED POSITION OF DENTAL THERAPEUTICS AND MATERIA MEDICA, AND WHOSE WISDOM HAS BECOME RIPENED BY REFLECTION AS WELL AS BY EXPERIENCE.

NOTE.

The time-honored practice of authors addressing their readers with some explanatory remarks, in the form of a preface, seems superfluous; the book should tell its own story and make its purpose clear. I therefore simply desire to make acknowledgment for the aid I have received in preparing this little work.

The following are the authorities consulted: "The American System of Dentistry," Litch; "Principles and Practice of Dentistry," Harris; "Dental Medicine," Gorgas; "Compend of Dentistry," M. Parreidt; "Dental Science," Ingersoll; "Chemistry," Leffmann; "Materia Medica," Potter; "Surgery," Gross; "Diseases of the Jaw," Heath; Dulles, "What to Do First;" and Butler's "Emergency Notes."

I am also indebted to Prof. C. N. Peirce for valuable suggestions and assistance in revising proof.

GEO. W. WARREN.

Lansdowne, Pa.
   January, 1893.

CONTENTS.

	PAGE
ANATOMICAL AND PHYSIOLOGICAL INTRODUCTION,	9
Development of the Teeth,	9
Structure of the Teeth,	12
Anatomy of the Teeth,	16
Decalcification of the Temporary Teeth.	21
Dental Pathology and Therapeutics,	25
Inflammation,	25
Diseases of the Dental Pulp and Membrane,	31
Diseases of the Hard Dental Structure,	37
Injuries and Diseases of the Maxillary Bones,	40
Defects of the Palatine Organs,	58
Extraction of Teeth,	61
Calcareous Deposits,	63
DENTAL MEDICINE,	68
Narcotics and Hypnotics,	68
Analgesics or Anodynes,	71
Anæsthetics,	74
Stimulants,	88
Tonics,	96
Sedatives,	110
Antipyretics,	110
Irritants,	111
Astringents,	114
Styptics and Hæmostatics,	117
Escharotics or Caustics,	118
Antizymotics (Antiseptics and Disinfectants),	121
Cathartics,	129
Emergencies,	130
WEIGHTS AND MEASURES,	156
INDEX,	159

ABBREVIATIONS.

ABBREVIATION.	LATIN.	ENGLISH.
āā,	Ana (G.),	Of each.
Ad saturand.,	Ad saturandum,	Until saturated.
Ad lib.,	Ad libitum,	At pleasure.
Aq.,	Aqua,	Water.
Aq. dest.,	Aqua destillata,	Distilled water.
Comp.,	Compositus,	Compound.
Ext.,	Extractum,	An extract.
F. or Ft.,	Fiat vel fiant,	Let there be made.
Garg.,	Gargarysma,	A gargle.
Gr.,	Granum vel grana,	A grain or grains.
Gtt.,	Gutta vel Guttæ,	A drop or drops.
Infus.,	Infusum,	An infusion.
M.,	Misce.,	Mix.
Mist.,	Mistura,	A mixture.
O.,	Octarius,	A pint.
Pil.,	Pilula vel pilulæ,	A pill or pills.
Pulv.,	Pulvis vel pulveres,	A powder or powders.
q. s.,	Quantum sufficit,	A sufficient quantity.
℞.,	Recipe,	Take.
S.,	Signa,	Write directions.
Spts.,	Spiritus,	Spirits.
ss.,	Semis,	The half.
Syr.,	Syrupus,	Syrup.
Tinct.,	Tinctura,	A tincture.
℔,	Libra,	A pound.
℥,	Uncia,	An ounce.
ʒ,	Drachma,	A drachm.
϶,	Scrupulus,	A scruple.
f℥,	Fluiduncia,	A fluid ounce.
fʒ,	Fluidrachma,	A fluid drachm.
♏,	Minim,	A drop.

DENTAL PATHOLOGY AND DENTAL MEDICINE.

ANATOMICAL AND PHYSIOLOGICAL INTRODUCTION.

DEVELOPMENT OF THE TEETH.

The **Enamel Organ.**—During the seventh week of fœtal life there appears on the border of the jaw a ridge of epithelium, known as the dental ridge; from this the epithelial follicles are deflected inward, and later on form what is known as the Enamel Organ. Each follicle represents an individual tooth. The cells active in the formation of the enamel are known as enamelblasts or amelloblasts.

The Dentine Organ.—A papilla arises in the dermal tissue, at a point immediately in contact with the rounded portion of the enamel organ—that is, from below in the lower jaw, and from above in the upper jaw. Simultaneously, the bottom of the enamel organ is rendered concave, in correspondence to the form and size of the dentine papilla, covering it like an inverted cup; this dentine bulb begins to assume the form of the tooth from the ninth to the tenth week. (See Fig. 1.)

By the end of the twelfth week, the follicles of the first or deciduous set of teeth are completed.

When the follicle is completed, it is developed by

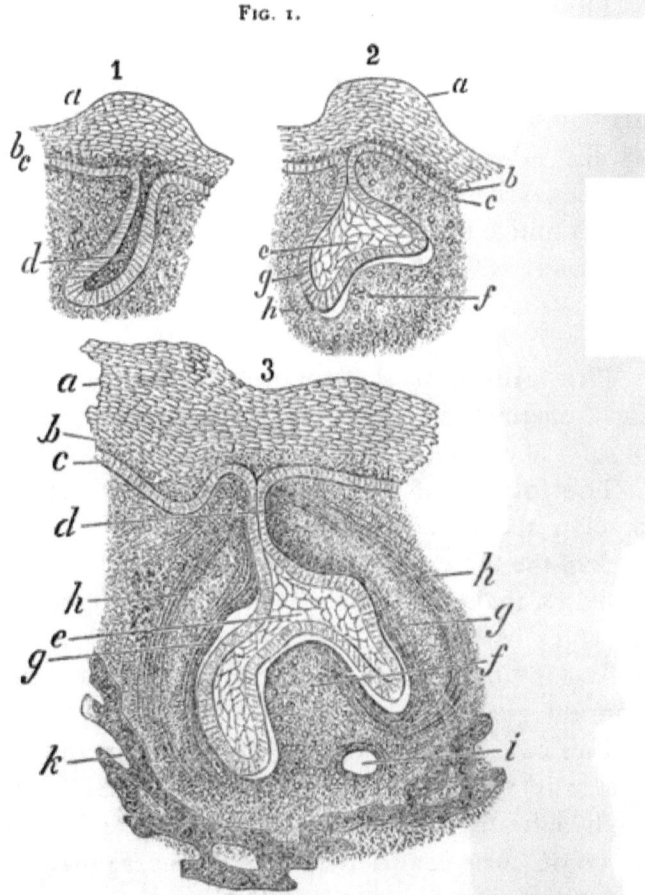

Fig. 1.

Three Stages in Developing Enamel Organs.

a, dental ridge; *c*, infant layer of cells; *d*, epithelial cord; *e*, stellate reticulum; *f*, dentinal papilla; *g*, inner tunic; *h*, outer tunic; *i*, transverse section of vessel; *k*, forming bone.

vascular tissue, forming the dental sack; this is about the fourteenth week. During the process of dentification there appears in the jawbone a groove within which the dentinal follicles rest.

The follicles of the permanent teeth begin to develop from the sixteenth to the seventeenth week of embryo life; of these, the first to become completed, are those of *the four first molars, at about the twentieth week;* these are followed by the follicles of the teeth anterior to them, which are completed before birth. These originate from the necks of the primitive follicles. *The bud of the follicle of the first permanent molar* originates from the epithelial lamina, as do the deciduous teeth, and back of all the follicles of the temporary set.

The follicle of the second permanent molar originates from the neck of the follicle of the first molar, and begins to form about the twelfth week after birth.

The follicle of the third permanent molar begins to develop about the third year after birth, budding from the neck of the follicle of the second molar, and is in its follicular evolution about three years, thus loosening its connection with the epithelial band at about the sixth year. It is then from twelve to fifteen years coming to such maturity as to emerge from the gum.

The calcification of the enamel (by the deposition of lime salts) commences at the surface of the dentine and proceeds outward, each enamel rod taking the form of the cell, which in their compactness become hexagonal, and the lime salts from each are deposited in immediate contact with each contiguous cell. When it is completed, the enamel organ disappears and nourishment to the enamel can only be had through the surface lying nearest to the dentine, which is accomplished through osmotic action.

The calcification of the dentine commences on the surface next to the enamel organ, and progresses from without inward, this being the point where mineralization begins in the tooth structure. The working cells of the dentine, the *odontoblasts*, throw out processes around which the lime salts are deposited, the processes lengthening with the thickening of the cap of dentine; thus forming the tubular structure of dentine. The processes occupying the tubules of the dentine are simply protoplasmic prolongations. As age advances, both the tubuli and fibrilli decrease in size, and in old age the extremities are nearly or wholly obliterated.

The Cementum.—The cells active in the formation of the Cementum are termed *Cementoblasts*. This structure is formed from the pericemental membrane, remaining as the residue of the dental sack, and becomes adherent to the previously calcified dentine.

Calcification commences about the seventeenth week of fœtal life, in the temporary incisors, and in the remaining temporary teeth during the seventh month; in the first permanent molars during the eighth month; during the first year in the permanent incisors and cuspids; the third year in the bicuspids, the fifth year in the second molars, and during the eighth year in the third molars.

It requires about two years for calcification to become completed in a deciduous tooth, and about ten years in a permanent tooth. (See Fig. 2.)

STRUCTURE OF THE TEETH.

Physiologically, the teeth are divided into the enamel, dentine, cementum, pulp, and pericemental membrane.

The Enamel covers the crown portion of the dentine. It is the hardest and most dense of all organic substances, hence it serves as a protection to the dentine from abrasion, and

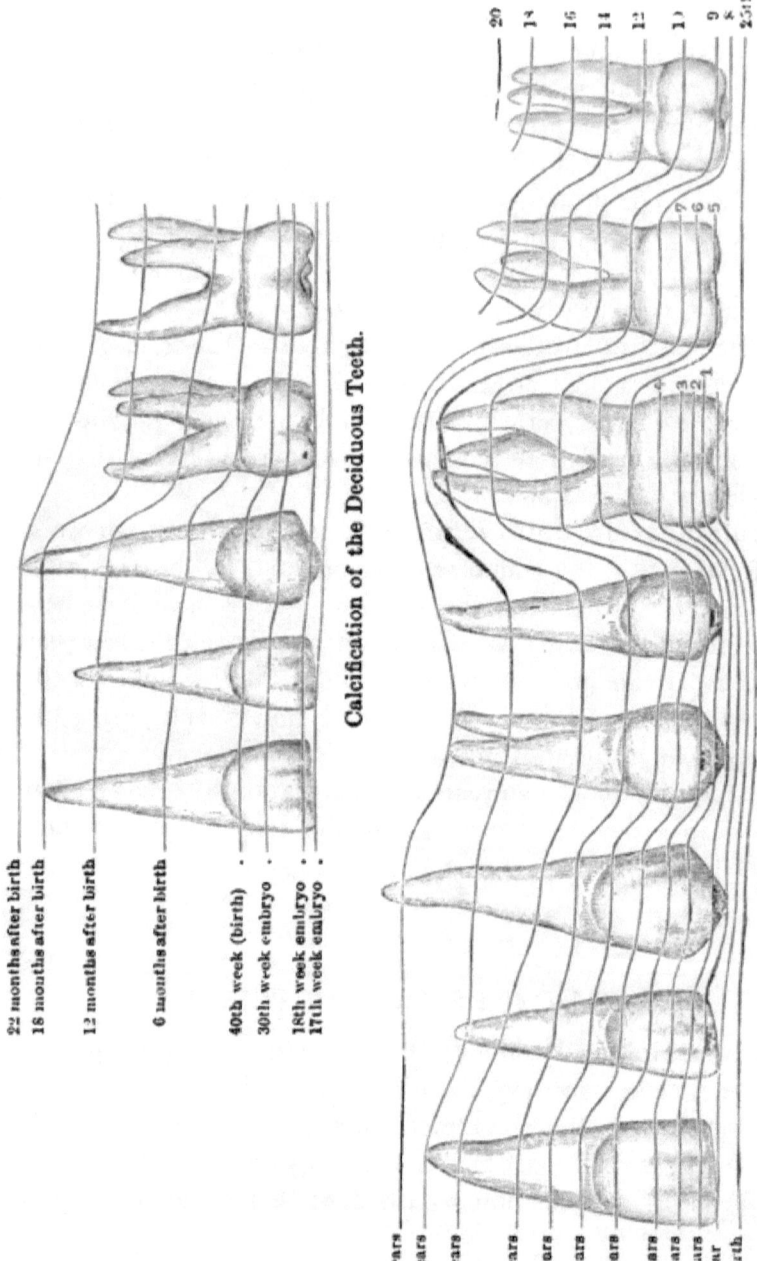

forms a greater resistance to mechanical force and the action of acids; it also serves to beautify the teeth.

The dentine gives the typical form of the teeth. The chief characteristics which adapt it to constitute the main body of the tooth structure are its density and vitality. While it encloses within it the pulp, it is itself enclosed on the crown by the enamel and on the root by the cementum.

Interglobular Spaces.—The imperfect dentine formation, known as interglobular spaces, found near the surface of the enamel, are due to mal-nutrition. Pits in the enamel will very often be found accompanying these interglobular spaces, and are due to the same causes.

The Cementum covers the root portion of the dentine. It is a somewhat dense substance. Its special use is, by being intermediate in the density of its structure, to form a union of the soft tissue of the root membrane with the dentine, thus aiding the pulp in nourishing the tooth, and preserving the vitality of the tooth after the pulp may have been devitalized.

The Pulp is enclosed by the dentine, and represents the shape of the tooth in a diminished size. It is composed of nerve, vein, artery, and formative tissue. This body is the mere shrunken condition to which the tooth germ, or dental papilla, is reduced, after it has normally accomplished the work of dentification.

The pulp is exquisitely sensitive and highly vascular; it is of a reddish-gray color and is enveloped in an exceedingly delicate membrane, which is continuous with the peridental membrane and is adherent to the wall of the pulp cavity.

The pulp is divided into two portions—the crown or body, which occupies the crown cavity, and the extremities, which occupy the root canal.

In advanced age this body often undergoes considerable

change; the size is sometimes much diminished, while again it is found as a shriveled and nearly insensitive mass.

This degeneration is due to the irritation and shock to which the teeth are liable at this time of life, owing to their

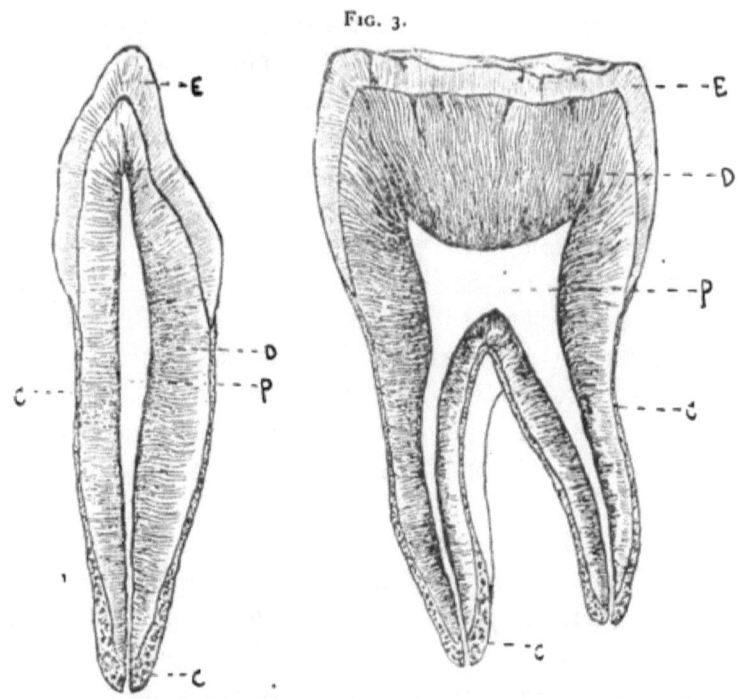

Fig. 3.

Represents Vertical Sections of Upper Cuspid and Molar Teeth, showing relative thickness of the hard dental tissues.

E.—Enamel.   D.—Dentine.
C.—Cement.   P.—Pulp Cavity.
(*After Sewell.*)

worn and abraded condition. If the irritation is mild and constant, and the patient of a calcic diathesis, a further deposit of lime salts is made, and the pulp diminished in size in consequence. But where the irritation is more severe it is apt to cause congestion and death of the pulp.

The **Pericemental Membrane** is, as the name implies, the membrane which surrounds or invests the roots of the teeth. It is a richly vascular, fibrous structure, and is the nutrient organ of the cementum. It is also the organ of touch in the tooth; by means of the nerves of the pericemental membrane, every touch upon the tooth is reported to the brain. It serves, too, to unite the tooth to the alveolus by its continuation throughout the alveolar cavities or sockets, and is connected at the dental foramen with the pulp, as spoken of above.

The peridental membrane also serves as a cushion, permitting a certain passive motion by which the teeth are protected from injury by blows and concussions which they are apt to receive during the performance of their peculiar function of tearing and grinding during the process of mastication.

ANATOMY OF THE TEETH.

Anatomically, the teeth are divided into three parts, the crown, neck, and root, the crown being that portion which projects freely into the mouth; the neck is surrounded by the gums, and the root is that portion covered by the alveolar process of the jaw and by which the whole tooth is held securely in position. In old age the gums recede, exposing the neck, and a part of the root is sometimes exposed in consequence of atrophy of the alveolar process.

The temporary denture consists of twenty teeth, divided into three groups—the incisors, cuspids, and molars. The relative position and number these groups bear to one another is expressed in the following formula:—

$$M_2 \ M_1 \ C \ I_2 \ I_1 \ | \ I_1 \ I_2 \ C \ M_1 \ M_2.$$
$$M_2 \ M_1 \ C \ I_2 \ I_1 \ | \ I_1 \ I_2 \ C \ M_1 \ M_2.$$

In the temporary teeth the proportion of the length to the width is marked, they being somewhat shorter than their successors, the permanent teeth. The color of the temporary teeth is of a milky or bluish white, while that of the permanent presents a yellowish appearance. In determining between the temporary and permanent teeth the observer will be aided by the relative size and color, and by remembering that the former are somewhat loose, and have a marked depression at the neck, just at the union of the enamel and cementum.

The permanent denture consists of thirty-two teeth, which are divided into four groups, namely, incisors (cutting), cuspids (tearing), bi-cuspids (crushing), and molars (grinding), according to the following formula:—

$$M_3\ M_2\ M_1\ BC_2\ BC_1\ C\ I_2\ I_1\ |\ I_1\ I_2\ C\ BC_1\ BC_2\ M_1\ M_2\ M_3.$$
$$M_3\ M_2\ M_1\ BC_2\ BC_1\ C\ I_2\ I_1\ |\ I_1\ I_2\ C\ BC_1\ BC_2\ M_1\ M_2\ M_3.$$

The relation between the permanent and temporary teeth is shown in the accompanying diagram, Fig. 4.

The anterior twenty teeth, namely, the incisors, cuspids and bicuspids, each have one root, except the first upper bicuspid, which in about eighty per cent. has two roots, one labial and one palatine. The roots of the upper incisors are rounded; those of the biscuspids are flattened laterally. The roots of the lower incisors are the most flattened, while the root of the cuspid combines partially the roundness of the incisor and the flatness of the bicuspid. The upper molars have three roots each, two buccal and one palatine; they are of a round shape. While the lower molars have but two roots, one anterior and one posterior, these are laterally flattened, and extend from the buccal to the lingual surface of the tooth. The roots of the third molar or wisdom tooth are subject to the same rules as those of other molars, but they are subject to a great number of exceptions.

The crowns of the teeth present several surfaces for examination, which are named according to their position and use. Those of the incisors and cuspids presenting toward the lips are called *labial surfaces;* while the same outer surfaces of the bicuspids and molars are called *buccal*, being next to the cheeks, and the opposite or inner surfaces of all the teeth, that portion presenting toward the tongue, are

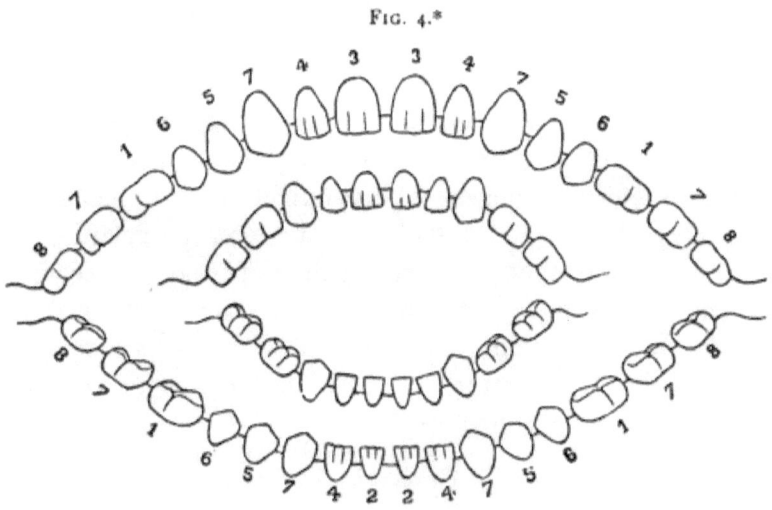

Fig. 4.*

The figures 1, 2, 3, etc., indicate the groups of teeth and the order of their appearance.

called the *lingual surfaces.* Some, however, use the term "palatine surfaces" for those of the upper jaw and "lingual" for those of the lower. While this is not necessary, it seems perfectly proper.

Those surfaces of the teeth that lie adjacent or next to the

---

*" Disease of the Digestive Organs in Infancy and Childhood," 2d ed. By Dr. Louis Starr.

adjoining teeth are called proximate or proximal ; these are more closely defined or divided by the terms *mesial* and *distal*. They are so named by their relative position to the central or median line of the face. Those proximate surfaces which face toward the median line are called *mesial surfaces;* and the opposite, or those most distant from this line, are called the *distal surfaces*.

The cutting edges of the anterior teeth and the grinding surfaces of the bicuspids and molars are called the *occluding* surfaces.

THE NORMAL ARTICULATION.

In the upper jaw the arch is larger than in the lower, especially in the anterior portion of it. The upper and lower teeth, therefore, do not meet perpendicularly, in articulating ; the lower incisors and cuspids articulate on the palatine surfaces of the corresponding teeth in the upper jaw. The cusps of the lower bicuspids and molars articulate in the grooves and furrows of the upper, and the teeth are so arranged in size and position that each tooth in the upper jaw when articulating occludes with two of the lower teeth.

DENTITION.

First Dentition takes place, normally, in the order given in the following formula :—

Central Incisors, ... 5th to 7th month.
Lateral   "   . . . . 9th to 11th   "
First Molars, ... 12th to 14th   "
Cuspids, ... 14th to 18th   "
Second Molars, ... 18th to 25th   "

Exceptions.—Some children are born with a few teeth erupted, and there are cases reported where the entire temporary set have been erupted at birth ; there are also cases on record of adults who have been edentulous from birth.

The force which causes the teeth to emerge is called *vital force*, and operates by growth, developing first the neck, then proceeding to the apical end of the root.

LESIONS INCIDENT TO FIRST DENTITION.

Though the eruptive process of dentition is a physiological action, it is very often attended by serious irritation. During early childhood the tissues are all softer, more vascular and sensitive, the nervous system predominating. Hence it is that the system at the period of first dentition is so susceptible to nervous impressions. In infancy, too, the system is less capable of combating diseases, and a large portion of the alarmingly great mortality of this period is traceable to the irritation caused by dental evolution.

The indications of the eruption of the teeth are, first, an increased flow of saliva—a healthy manifestation, as it tends to keep the mouth moist and cool. This "drooling" is due to the irritation of the trifacial or fifth pair of nerves, which is sensory to the teeth and nutrient to the salivary glands. When the irritation becomes more pronounced, the mouth becomes hot and dry, the cheeks unusually red, eruptions appear upon the face, and, indeed, sometimes over the whole body; with ulcerations upon the tongue and mucous membrane of the mouth and inflammation of the gum over the advancing tooth or teeth. A condition quite opposite to the above in appearance is a *white* and *hardened* gum overlying the advancing tooth or teeth. This offers a greater resistance than the inflamed condition and is often neglected, especially by physicians, owing to the absence of any appearance of congestion.

The child becomes wakeful, peevish, and cross, loses his appetite, and if relief is not then given this may be followed by persistent nausea, diarrhœa, and spasms or convulsions.

Treatment.—The first thing is to remove the irritation by freely lancing the gums over the advancing teeth. The manner of doing this is admirably and fully shown in the accompanying illustration (Fig. 5), which is taken from a paper by the late Dr. James W. White, in the "American System of Dentistry." If the convulsive stage is reached, the patient's feet should be placed in hot mustard water, and cold cloths applied to the head, or the entire body put in a warm bath; such measures cause muscular relaxation and have a soothing effect upon the nervous system.

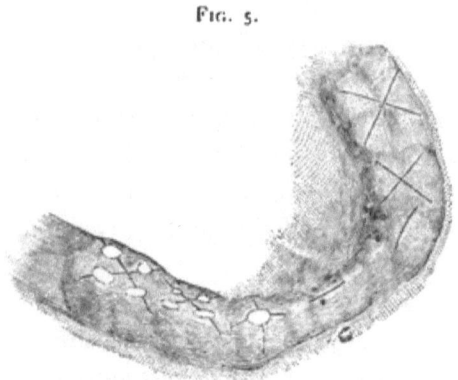

Fig. 5.

The temporary teeth must be removed before the permanent can erupt in their normal position; this takes place normally, by the absorption or decalcification of the roots.

Decalcification of the Temporary Teeth.

Decalcification of the roots of the deciduous teeth usually commences at the apex of the root, on the side nearest to the successional tooth; this, however, is not invariably the case. Absorption may commence at several and distinct points, sometimes on the labial side,—that most distant from the succeeding tooth.

The cause and manner in which the roots of the temporary teeth are absorbed has been the subject of much and careful study by such advanced investigators in the domain of dental science as Tomes, Peirce, Bodeker, Abbot and others.

It was thought for a long while that the pressure caused by the advance of the permanent teeth was the sole cause for the decalcification of the primary teeth; but it is now generally conceded that *it is simply the result of a physiological action* and not a mechanical force. The fact that the decalcification of the deciduous teeth is sometimes successfully accomplished in the absence of the corresponding permanent teeth adds much to the evidence that their presence and pressure is not essential.

Prof. Peirce, in the "Transactions of the American Dental Association," says: "The several conditions which are always present and essential are readily recognized, but the part each plays is not easily ascertained. The place of its commencement, at the end of the root, the retention of pulp vitality, and the presence of a vascular papilla in close proximity to the absorbing surface, with the fact that the surface of this papilla is rich in giant-cells, termed '*osteoclasts*,' are evidently essential accompaniments, and the absence of any one of them would certainly militate against the completion of the process."

And in another place the same writer says: "That the organ has served its purpose, and that the nourishment which had previously been appropriated by it is diverted or relegated to its successors, is probably the most plausible explanation we can give of this interesting physiological process."

Decalcification of the deciduous teeth commences in the central incisors at about the fourth year; in the lateral incisors in the fifth year; in the first molars the seventh year; the second molars the eighth year, and the cuspids the ninth year. After decalcification commences in a tooth, it takes about three years for it to accomplish its work. (See Fig. 6.)

Second Dentition.—In a harmonious development of the teeth and jaws, the indications of the time approaching for the development of the permanent teeth are the expand-

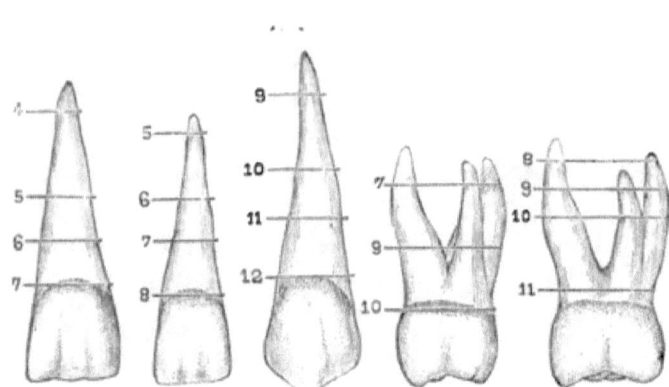

Fig. 6.

Decalcification of the Deciduous Teeth.

ing of the alveolar ridge and the spreading apart of the deciduous teeth.

The emergence of the permanent teeth takes place, normally, in the order given below:—

First molars,	6 to 7 years
Central incisors,	7 to 9 "
Lateral,	8 to 10 "
First bicuspids,	10 to 11 "
Second "	11 to 12 "
Cuspids,	12 to 13 "
Second molars,	12 to 13 "
Third "	17 to 25 "

Deviations from the order of appearance and the respective ages as given above occur; it is usually in strong children that the teeth appear at a later period, a scrofulous diathesis being indicated in premature dentition.

Third Dentition.—Cases of third dentition are reported by a number of writers, but undoubtedly the majority of these reports refer to teeth that in reality were part of the second set, which failed to emerge at the normal time, and had only appeared in old age when there was sufficient room for them, or the jaws had atrophied.

The writer has extracted two well-developed third molar teeth (which had never emerged from the gums) for a patient sixty years of age; this was after the extraction of the second molar roots. It was known that the patient never had had any "wisdom teeth."

Anomalies of Number.—Supernumerary teeth are especially frequent in the anterior portion of the superior maxillary. When sufficient room is wanted for them, they may stand within or without the arch. Usually the shape of both the crown and root of these teeth is conical, while the shape of the crown of those found in the posterior part of the mouth is cuboidal—resembling the molars.

Dentures from which normal teeth are missing are more frequent than those containing supernumerary teeth. Occasionally the space from which the permanent tooth is missing may be occupied by a temporary tooth. Sometimes temporary teeth may be seen in the mouth of persons twenty-five and thirty years of age; in these cases the permanent ones are generally retained in the jaw, and may erupt later.

DENTAL PATHOLOGY AND THERAPEUTICS.

Dental pathology treats of the origin and progress of the various diseases to which the teeth and the surrounding oral tissues are liable.

Dental therapeutics considers the medicines and remedies used in the treatment of such diseases.

Disease, pathologically considered, is any abnormal condition in those processes which constitute perfect health. It is but the normal physiological force perverted, tearing down what it had before built up. This cannot always be recognized at first, as it may be confined to an area so small as to involve but a single cell. Its chief manifestations are expressed by the word *inflammation*.

INFLAMMATION.

The most commonly observed features of inflammation are heat, redness, swelling, pain, and impaired function, modified only by peculiarities of the structure and intensity of action. It is a series of changing conditions, each the result of the preceding one.

The first cause is irritation; that is, the slightest foreign influence disturbing the harmony of the functions of the tissues. It is most readily traced in the vascular system. In this, the first stage of inflammation, the capillary vessels are excited into active contraction and relaxation, quickening the circulation of the blood, inducing warmth, and causing some discomfort.

Acute inflammation, or the second stage, is characterized by warmth, a sense of fullness, slight swelling, and increasing pain; the abnormal volume of blood in the parts presses upon the nerve filaments supplying the inflamed tissue, causing pain in accordance with the expansion of the vessels.

Chronic Inflammation.—In prolonged inflammation, the functions of the parts become somewhat changed, adapting themselves to the surrounding conditions; thus, the tissues involved become less susceptible to the impression of an irritant, and comparatively little pain follows, this being one of the most noticeable characteristics of chronic disease.

Treatment.—In treating inflammation effort is made to effect resolution, or to hasten suppuration.

Resolution is the subsidence of inflammation and a return of the tissues to normality.

Suppuration, or pus formation, being the breaking down of the parts, the debris of which, with white blood corpuscles, forms *pus*.

The remedies used are, first, to remove the cause, that is, the irritant, to administer laxatives, and apply sedatives and stimulants.

The chronic stage of the inflammatory process, if not checked, may result in hypertrophy, induration, tumefaction, or congestion of the tissue.

Hypertrophy.—Hypertrophy is an excessive growth of normal tissue by the multiplication of cell elements. Of the oral tissues, the gum and mucous membrane are the most liable to this disease. The usual form recognized is in a growth of the free border of the gum, lying loosely against the teeth, in some cases extending to the cutting edges. Another form consists in the thickening of the entire gums, covering the teeth to such an extent that only the masticating surfaces are visible; the hypertrophied portion being firm and dense, protruding the lip to such an extent as to cause deformity; this latter form of hypertrophy of the gum, however, is rare.

Treatment consists in cutting away the long points and borders of the gum down to the necks of the teeth, and then reducing the inflammation with stimulants and astringents.

Induration is a circumscribed, hardened swelling; it is an enlargement of individual cells, and not a multiplication of the cell elements, as in hypertrophy. In this stage of chronic inflammation the functions of the tissue involved are inactive, the circulation being very much retarded. It may manifest itself in the gum, in the muscles of mastication, or in the muscles of the neck.

Treatment.—In induration of cheek or neck, apply hot cloths and an active stimulant, such as capsicum. When the disease is in the gum, apply capsicum plaster and lance freely.

Tumefaction is a condition of chronic inflammation, by which is produced an abnormal amount of tissue, of a *different* kind from the surrounding tissue, exhibiting a difference in color and texture. Tumefaction as found in the mouth is divided into *epulis, cystic, and vascular tumors*.

Epulis tumors originate in the periosteum, and are usually found in the interspaces of the anterior teeth. They are fibrous in structure, and usually of a dark red color.

Treatment consists of excision, care being taken to remove all the abnormal tissue; the bistoury should be passed well around and under the tumor, as deep as the periosteum; after removing the growth as entire as possible with the knife, carbolic acid should be applied, to destroy the vitality of any fibres that should remain.

Cystic tumors originate in the mucous membrane; they are membranes in structure, of a lighter color than the normal mucous membrane, and are filled with a viscid fluid.

Treatment consists in lancing at the lowest points, and emptying the contents by pressure, after which applications of stimulants and astringents should be made.

Vascular tumors arise from some blood-vessel supplying the oral tissues. They are usually of a dark red color, and highly vascular; they have a smooth and shiny surface, and are more or less fibrous.

Treatment same as in epulis tumors.

Congestion is often the result of the inflammatory processes.

The prominent feature of congestion is *blood stasis*, causing severe pain with each pulsation of the heart. The capillary vessels in the inflamed parts being engorged, the blood rebounds at each pulsation of the heart, thus causing the throbbing pain. The pain, however, is sometimes intermittent, by the capillaries being broken through, relieving the pressure temporarily.

Treatment is to apply stimulants and sedatives.

Alveolar Abscess.

An abscess is a circumscribed cavity, containing pus—circumscribed by the formation of a soft membrane, forming a sack, restraining the pus from passing into the surrounding tissues, causing it to seek the surface at the point of the least resistance, here parting the tissues, forming an escape, called a *fistulous opening*.

Alveolar abscess is, as the name indicates, an abscess originating within the alveolar walls. It is the result of inflammation of the pericemental membrane, consequent, usually, upon the death and septic decomposition of the tooth pulp.

Symptoms of alveolar abscess are congestion and inflammation of the gums about the affected tooth; severe pain which is often accompanied by considerable fever. There is also apparent elongation of the tooth, caused by inflammation of the pericemental membrane and accumulation of pus in its socket; the pus formed, being confined in the apical space between bony walls, causes considerable pressure as it accumulates. The bone in this neighborhood is not of a very hard nature, and in consequence is readily destroyed by the pus in its effort to seek the surface. During the time the pus is penetrating the bone is when the patient suffers most; the pain at this stage is very severe, and of a throbbing character.

The features become swollen and disfigured on the affected side; the eye is sometimes entirely closed, and the jaw so stiff that the mouth cannot be opened to any considerable distance.

Other causes than that of the **death** and decomposition of the pulp may give rise to alveolar abscess : any foreign matter, such as *filling material*, being forced through the apex of the root, *calcic deposits* within the walls of the alveolus, *necrosed root or bone, impacted teeth, etc.*, may cause sufficient inflammation of the periosteum to produce an abscess, all the successive stages of inflammation being involved in the formation of an abscess, from irritation to suppuration.

Treatment.—The surgical and local treatment consists in gaining free access to the diseased parts, and removing the cause, and breaking up the sac of the abscess. In securing ready access to the point of accumulation it is best to first open up the canals of the affected tooth. In doing this it is better to sacrifice good tooth structure than to attempt to work around corners and through too small a canal, being careful not to go through the side of the root.

Where the abscess has advanced to any considerable extent, it will often be found necessary to more freely open or enlarge the fistulous canal, by drilling in through the gum and alveolar process over the affected tooth with a medium sized rose-head burr revolved by the dental engine. In so doing the parts may be more thoroughly drained and treated, and the sac, when attached to the end of the root, can be more readily reached and broken up. If the patient is in an otherwise healthy condition, nature will finish the work, by throwing off the broken-down tissue and developing new granulations, without further treatment. It would be advisable, however, to assist in throwing off the pus and other foreign matter by injecting peroxide of hydrogen freely into the canals and socket.

Where the patient is of low vitality, or the abscess of long standing, other therapeutic treatment must follow the surgical before a cure can be accomplished. First cleanse the canals thoroughly with peroxide of hydrogen, as before stated, and then some one of the more efficient antiseptic and disinfectant remedies, such as bichloride of mercury, carbolic acid and iodine or aristol should be applied on a strand of floss silk, or a few fibres of cotton, which should be carried to the apex of the root by means of a nerve instrument, and the crown cavity closed with some one of the temporary stoppings. The gums about the affected tooth should then be painted with tincture of iodine and aconite, which will aid in the reduction of the inflammation by counter-irritation. This treatment should be repeated in from two to three days, and continued according to the character and symptoms of the case in hand.

ULCERATION.

Ulceration is an open *suppurating surface*. When ulceration is at all deep-seated, it differs from an abscess in not having a lining membrane, forming a sack, but involves all contiguous structure, the pus passing through the tissue at one or several partings.

Treatment consists in applying astringents and antiseptics, and *administering* a tonic.

Pyorrhea Alveolaris.

The most **remarkable feature** of this disease is the chronic discharge of pus around the necks of the teeth. *There is a deposit* on the roots, either sanguinary or salivary, or both, signifying an excess of lime salts in the system. *The gum* appears livid, and a pocket is formed on the root of the tooth, in which the pus accumulates. As the disease progresses the pockets become deeper and broader, the quantity of pus discharged increasing proportionately. This may appear about one tooth or a number of them at the same time.

The **alveoli** becomes involved, the gums gradually recede, exposing the roots of the teeth, which are soon loosened and lost in consequence.

Treatment.—It may be so modified or retarded that the teeth may be kept in a useful condition for several years. When it appears in the mouth of a person under twenty-five years of age (which is rare) a cure is possible. In the majority of cases, however, it does not make its appearance until thirty-five or forty years of age, when a permanent cure is doubtful.

The remedies usually employed, and that have proved most successful, are as follows: All salivary and sanguinary deposits are removed; the pockets are then thoroughly cleansed, from two to three times a week, with peroxide of hydrogen, followed by an application of aromatic sulphuric acid or chloroform, and aristol as a disinfectant and stimulant.

DISEASES OF THE DENTAL PULP AND MEMBRANE.

Inflammation of the Pulp.—As in other tissues, inflammation of the pulp is induced by irritation. *The most frequent cause* is the encroachment of the dental caries; this removes the normal covering of the pulp, allowing thermal, chemical, and mechanical irritation to readily reach that soft tissue.

Treatment.—If irritation has been mild and only for a few days, immediate filling will give relief. *If the irritation* continues for several days and has been severe enough to cause any actual pain, *treatment consists* in applying some stimulating antiseptic and filling with some good temporary stopping. If no further pain is felt within ten days, fill permanently. *When the irritation* has continued for a considerable length of time, the pulp being seriously involved though it may not be exposed, or in cases where the pulp is exposed from the disintegration of the dentine covering, and is subject to paroxysmal pains, it is generally preferred to devitalize and extirpate the pulp before attempting to fill the cavity.

In teeth that are not carious, inflammation of the pulp may arise from violent thermal changes, to which they are very sensitive, and if repeatedly exposed to such changes, death of the entire pulp may be the result. Hence we not infrequently find teeth with devitalized pulps that are otherwise sound.

Treatment in such cases is very simple. It consists in removing the irritant; that is, protect the teeth from such marked thermal changes, and apply some good stimulant, such as tincture of iodine and aconite, to the gum over the root of the tooth affected.

Inflammation of Pulp following Filling.—Occasionally inflammation of the pulp follows the filling of carious teeth. This may arise from either *mechanical* irritation, that is, undue pressure upon the pulp by the capping or filling material; *chemical*, by the action of acids used in the capping or filling material; or *thermal*, by the conductibility of the filling material, readily carrying every impression to the pulp.

Treatment.—This can be readily diagnosed, and when the inflammation is only superficial the pulp may be restored to health by removing the irritant and treating as directed above. But when inflammation is more general (and every precaution has been taken in filling the tooth), there is no reasonable probability that it can be saved ; the filling, therefore, should be taken out, the pulp destroyed and removed, and the tooth refilled.

Odontalgia (tooth pain) is not infrequently caused by sensitive dentine on the abraded masticating surfaces of the teeth. The enamel covering being removed, the dentine more readily carries the irritation to the pulp. If the irritasion is very severe, and is continued for a considerable length of time, it will cause *acute inflammation* and lead to the death of the pulp.

Treatment.—Apply to the abraded surface chloride of zinc crystals, or touch with nitrate of silver, and follow, when practicable, by building up the surface with gold ; also make application of tincture of iodine to gum.

Hypertrophy of the pulp is preceded by chronic inflammation. *It is the result of caries.* The nerves do not retain the same degree of growth as the connective tissue of the pulp ; the pulp polypus is, therefore, not so painful as the normal or acutely inflamed portion ; for this reason such growths are permitted to remain in the cavity for months before the patient calls for treatment.

Treatment consists in the extirpation of the polypus, then the devitalization and removal of the remainder of the pulp, before filling.

Pulp nodules or nodular dentine is a formation of small nodules of calcified matter within the pulp cavity ; they are generally confined to the body of the pulp, but at times nodules are found within the root canals. Since nodular calcifications sometimes occur in the pulps of teeth the crowns of which are perfect, there being neither abrasion nor decay, though decay may be, and often is, found in connection with such formations, the presence of these bodies is evidently due to some other cause that is as yet unknown. These bodies give much trouble when connected with other diseases affecting the tooth pulp, but *pain from this cause alone is very rare.* When these bodies do occasion trouble, it is generally in the form of infra-orbital neuralgia, with paroxysms of pain in one or more of certain teeth. Patients with these symptoms sometimes present themselves. In these instances, where no carious or otherwise visibly diseased teeth are found, it may be presumed that the pain is caused by an odontinoid formation in the pulp of one of these teeth. To ascertain the affected tooth, the usual tests of cold water and percussion should be employed, when the patient will usually express some abnormal feeling in the tooth or teeth containing these formations.

Treatment.—When pulp stones are diagnosed, the tooth should be drilled into in the best position and direction for reaching the pulp; this done, arsenious acid should be applied for the devitalization of the pulp, which, of course, should be thoroughly removed and the root canals and tooth filled.

Gangrene of the Pulp (death in a body without loss of substance, as in suppuration).—Gangrene of the pulp not infrequently is the result of acute inflammation; the over supply of blood in the arteries compresses the veins at the dental foramen, almost or entirely interrupting the circulation, thus causing the death of the pulp *in a body*. *Treatment* consists in the removal of all remnants of pulp and in perfect disinfection of the root canals before filling.

Pericementitis (inflammation of the pericemental membrane) is consequent upon some one or more of the following irritants: inflammation of the pulp, putrescent pulp, excess of filling material, looseness of the tooth or root, salivary or sanguinary calculus, dental manipulation, mal-occlusion, want of occlusion, mercurial poisons, previous pericementitis, etc.

Symptoms are knowledge of the presence of the tooth, apparent elongation of the tooth, pain following pressure from occlusion of the jaws, from the tongue, fingers, etc., the most decisive test being the tapping of the tooth, which, under these conditions, is followed by severe pain.

Treatment.—First remove the irritants, vital or mechanical, then apply a good stimulant and sedative to the gum over the affected tooth, to hasten resolution. A hot foot-bath is also beneficial in relieving the blood pressure in the diseased parts. When gangrenous or putrescent pulp is diagnosed, the teeth should be thoroughly opened up, every vestige of the pulp removed, the root canals thoroughly cleansed and disinfected as soon as the soreness will permit. "The ancient plan of drilling a vent-hole, to relieve a tooth from the pressure of the gases forming within the pulp chamber, should be consigned to past history." *

Exostosis.

Exostosis, or hypercementosis, is a disease common to all bones, but owing to the vascularity of the cementum of the tooth it is oftener found there to a greater or less degree than in any other part of the osseous structure. It consists of outgrowth of new tissue from the cemental layer covering the roots of the teeth. It sometimes takes the form of prominent nodules, and again will be found in regular layers or masses covering a large portion of the cementum.

The cause of exostosis is inflammation of the pericemen-

---

* Prof. G. V. Black.

tal membrane (pericementitis), which may be induced by mal-occlusion, want of occlusion, shock from severe dental operations, or other violence, such as biting thread—a habit formed by many seamstresses. Harris, in his "Principles and Practice," does not agree with this theory, that hypertrophy of the cementum is attributable to irritation of the peridental membrane, but states (page 354) "that it seems to be due to some constitutional diathesis." This we cannot entertain without some demonstration; at most it can only be a predisposing cause. Where any of the above conditions are continued for any length of time it causes an abnormal energy in the odontoblastic layer of the cementum, producing increase in the growth of the structure. The alveolus is in many cases enlarged or absorbed in proportion to the growth of cementum, and the patient experiences little or no inconvenience, but there are often instances where the enlargement of the cementum causes such pressure on the nerves as to give more or less discomfort, and sometimes excruciating pain; at times, too, it causes severe facial neuralgia quite remote from the seat of the trouble. It also happens in this way that the roots of adjacent teeth sometimes become firmly united.

Treatment.—If it is possible to discover this disease at an early stage, the application of a good counter-irritant, such as tincture of iodine, over the affected root may interfere with its progress. But where the disease has established itself the extraction of the affected tooth or teeth is the only available treatment.

DISEASES OF THE HARD DENTAL STRUCTURE.

The Etiology of Dental Caries.—Dental caries is the gradual softening and disintegration of the tooth substance. It first appears as a chalky, opaque spot in the enamel, in which the structure is loosened and gradually broken down. The primary cause is unquestionably the product of fermentation (acid) of particles of food, and the abnormal thickened mucus on the surfaces of the teeth.

The proximal surfaces are convenient positions for the lodgment of fermentable substances; in consequence, fully 65 per cent of the first appearances of caries occur on these surfaces of the teeth. From the statistical examinations made by the writer of one hundred thousand permanent teeth, dental caries may be classified in the following groups:—

	Per Cent.
Superior central incisors, ... carious	.144
Inferior " " . . . . "	.006
Superior lateral " . . . . "	.14
Inferior " " . . . . "	.008
Superior cuspids, ... "	.07
Inferior " . . . . "	.009
Superior first bicuspids, ... "	.10
Inferior " " . . . . "	.03
Superior second bicuspids, ... "	.08
Inferior " " . . . . "	.05
Superior first molar, ... "	.16
Inferior " " . . . . "	.145
Superior second molar, ... "	.11
Inferior " " . . . . "	.12
Superior third " . . . . "	.025
Inferior " " . . . . "	.03

Relative Location of Dental Caries.

	Approx.	La.	Pal.	Mas.	Buc.
Superior central incisors,	.95	.01	.04	. .	. .
" lateral "	.94	.01	.05	. .	. .
" cuspids	.98	.015	.005	. .	. ?
" first bicuspids,	.93	.005	. .	.065	. .
" second "	.92	.005	. .	.075	. .
" first molars,	.35	. .	. .	.61	.04
" second "	.20	. .	. .	.75	.05
" third "	.04	. .	. .	.95	.01

	Approx.	La.	Lin.	Mas.	Buc.
Inferior central incisors,	.99	.01	. .	. .	. .
" lateral "	.98	.02	. .	. .	. .
" cuspids	.95	.045	.005	. .	. .
" first bicuspids,	.91	.01	. .	.08	. .
" second "	.91	.005	. .	.085	. .
" first molars,	.30	. .	. .	.64	.06
" second "	.16	. .	. .	.78	.06
" third "	.02	. .	. .	.95	.03

NOTE.—The above tables are made without respect to age or sex.

Dental Caries During Illness.—In severe illness, several conditions favor dental caries: 1st, the lack of nutrition causes the teeth to be less able to resist destructive influences; 2d, the abnormal acid secretion aids in the destruction of the tooth substance; 3d, fermentation often progresses about the tooth without hindrance, on account of the patient's inability to keep the surfaces properly cleansed.

Dental Caries in its Relation to Sex.—In the male adult the conditions are more favorable than in the female. The male usually uses his teeth more in mastication, and partakes of sweets less frequently between meals, there being a less desire for saccharine and farinaceous foods. Particles of these collecting about the teeth are readily converted into lactic acid, thus becoming injurious to the tooth substance. "The female is also at a disadvantage during pregnancy, at which time a large quantity of lime phosphates are essential to the growth of the fœtus, and this supply is diverted from the teeth. It is well known that during pregnancy fractures heal less readily, because the lime phosphates are needed for the fœtus. If, in a similar manner, the nourishment of the teeth is affected, as during the healing of fractures, one may readily conceive that their power of resistance against unhealthy influences must be materially diminished, and the reactions of the oral fluids during pregnancy is not infrequently acid. Finally, in these cases, the reflex disturbances of digestion should be considered also, for the acid eructations are also injurious to the teeth."*

Therapeutics of Dental Caries.—The most effective treatment of caries consists in the removal of the diseased portions and the proper preparations of the cavity, followed by a well inserted filling of the least destructible substance compatible to the tooth substance. There are exceptional cases, however, where decay does not extend into the dentine—where the tooth may be preserved by removing the diseased parts with a fine stone, diamond disk or diamond point, and leaving the surface thus treated polished perfectly smooth.

Prophylaxis.—From the time of the eruption of the deciduous teeth the mouth should be kept scrupulously clean. A soft or medium brush and water should be employed daily after each meal, and with a good powder upon rising in the morning and just before retiring at night. Where the teeth are so closely situated that the brush cannot be worked between them, so as to cleanse the proximal

---

* M. Parreidt.

surfaces, floss silk or quill toothpicks should be used to remove any particles of food that become lodged between the teeth.

Mouth Washes.—Alkaline mouth washes, such as lime-water, borax, or bicarbonate of soda in solution, are sometimes used ; not, however, as a substitute for the tooth brush, but as an adjunct. They neutralize the oral fluids when they are acid.

The best antiseptic mouth wash is bichloride of mercury, from 1 to 5000 to 1 to 15,000. These are not necessary, however, when the tooth-brush is properly employed.

INJURIES AND DISEASES OF THE MAXILLARY BONES.

Fractures of the lower jaw are usually the result of direct violence, such as a kick from a horse, fall from a height upon the face, the unskillful application of the dental "key" and forceps, etc. Professor Pancoast, however, met with a case in which the neck of the bone was fractured by a violent fit of coughing. The patient was nearly seventy years of age.*

The Most Frequent Location.—Fractures occur most frequently in the neighborhood of the cuspid tooth, this position being determined by the weakness of the bone at this point, in consequence of the depth of the alveoli.

Fractures of the alveolus are frequently unavoidable during the extraction of teeth. The displacement of portions of this bone, however, gives little inconvenience, and hastens the absorptive process. Should the fracture affect the alveoli of the adjoining teeth, a troublesome exfoliation may follow.

---

* Gross's "Surgery."

Since accidents of this kind are due to the natural conformation of the parts, legal proceedings against the operator for this mishap are most unjust.

Diagnosis.—Fractures of the lower jaw are readily recognized; the regularity of the dental arch is altered, and the mobility of the fractured portions is shown when pressure is applied to the teeth or alveolar process at the site of pain.

Crepitation is discernible during the first week after fracture. Its absence, after this time, is due to formations of granulations and of the partial union of the fractured ends.

The gums, also, are usually lacerated at the point of fracture, accompanied by considerable inflammation and swelling.

Fracture of the ramus of the jaw is less frequent than in the body, and is not so readily diagnosed, as the upper portion cannot be grasped with the fingers, and crepitation is difficult to make out.

Treatment of Fractures of the Lower Jaw.

After the reduction of any displacement, the treatment is usually of a simple character, though there is a case occasionally where the most carefully adapted mechanical appliances fail to effect a good union.

The appliances used for the maintenance of the fractured portions in position may be divided into two classes:—

External and internal to the mouth, though it may be, in a few instances, necessary to combine the two methods. The simplest form of apparatus for external use is the ordinary four-tailed bandage or sling, with a slit for the chin to rest in (Fig. 7). It is made from a piece of muslin, about a yard in length and two or three inches broad; this should have "a slit four inches long cut in the centre of it, parallel to, and an inch from, the edge. The end of the bandage should then be split to within a couple of inches of the slit, thus forming a four-tailed bandage, with a hole in the middle. The central slit can be readily adapted to the chin, the narrow portion going in front of the lower lip, and the broader beneath the jaw, and the two tails corresponding to the lower part of the bandage are then to be carried over the top of the head, while the others are crossed over them and

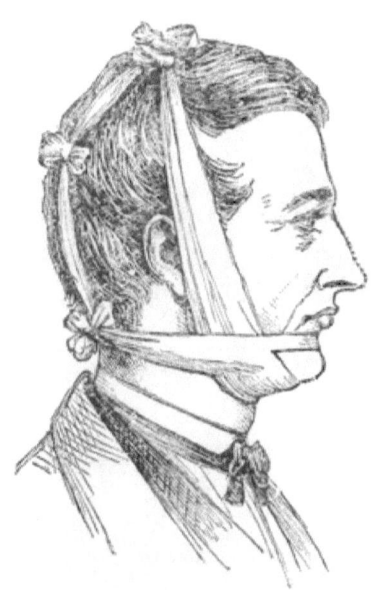

FIG. 7.

tied round the nape of the neck. The ends of the two bandages may then be knotted together." *

Hamilton has devised a sling for which he claims superiority; we give it in his own words: "The advantage of this dressing over any which I have yet seen consists in its capability to lift the anterior fragment vertically, and at the

---

* Heath.

same time it is in no danger of falling forward and downward upon the forehead. If, as in the case of most other dressings, the occipital stay had its attachment opposite the chin, its effect would be to draw the central fragment backward. By using a firm piece of leather as a maxillary band, and

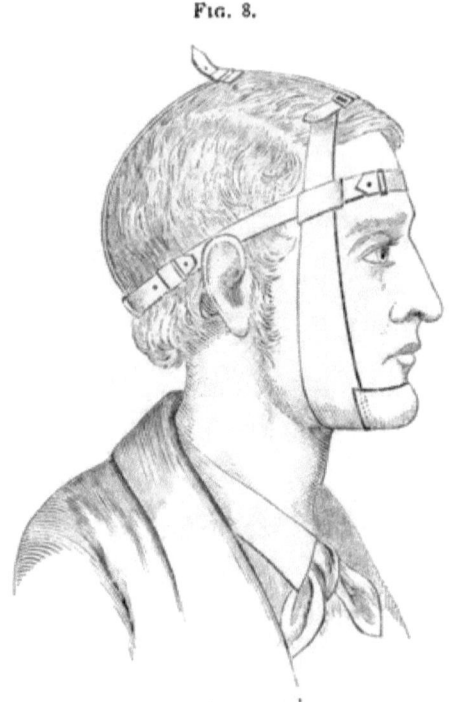

FIG. 8.

attaching the occipital stay above the ears, this difficulty is completely obviated" (see Fig. 8).

The interdental splint is an apparatus used in common among dentists for an internal appliance in the treatment of a fractured jaw. It is usually made of vulcanite rubber, and gives very satisfactory results. In 1866, Mr. Gunning, of

New York, gave a description of this contrivance as then used by him.*

Method of Making Single Interdental Splint.—Take impression in wax or modeling compound, using as small an amount as will insure a good impression of the teeth and gums. An assistant should stand behind the patient and hold the broken bone in place, while the operator stands directly in front and takes impression. . Cover the cast with (No. 60) tinfoil; this makes the cavities in the splint a little larger than the corresponding teeth, making it easily adjusted, and leaves it with a smooth surface. Use two thicknesses of base-plate

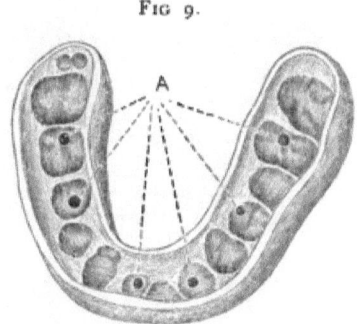

FIG 9.

wax over the tinfoil, allowing it to pass down a trifle below the necks of the teeth. Flask and vulcanize in usual manner for rubber work.

Fig. 9 represents such a splint, showing the inner surface which encloses the teeth and a part of the gum of the lower jaw. The holes marked A which extend through the splint are for the purpose of introducing a syringe in cleaning and treating the parts while the splint is in position. When proper treatment is employed, a cure is effected in from three

---

* *New York Medical Journal* and *British Journal of Dental Science.*

to eight weeks, according to the age and physical condition of the patient. When the fracture is of an obstinate vertical nature, a splint that will enclose the teeth and gums of the upper jaw as well as the lower should be used (see Fig. 10).

Fractures of the upper jaw are less frequent and less difficult of treatment than those of the lower jaw. In recent cases a simple replacement of the parts is all that is necessary, and occasionally the application of a simple retention splint is employed.

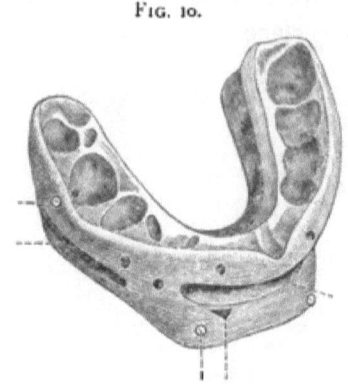

FIG. 10.

Angle's Method of Fixation.—A more recent and excellent method for treatment of fractures of the maxillæ is that devised by Dr. E. H. Angle, of Minneapolis. And through the Doctor's courtesy I am enabled to give a full description of his methods and appliances here. He says: "The methods used by myself in treating fractures of the maxillæ have been so successful and so gratifying that it would seem they approach for efficiency and simplicity more nearly the ideal than any yet devised."

In order that this system of treating fractures of the maxillary bones may be more easily understood, we will divide them into three classes. The first class comprises all simple fractures in which the teeth are good and sufficiently firm in their attachments (especially on each side of the fracture) to afford anchorage for the appliance.

The second class comprises all fractures where the teeth are unsuited, from disease or any other cause, for anchorage, but yet sufficient to give the correct articulation of the jaws.

The third class comprises all fractures where the jaws are edentulous. The following cases treated by Dr. Angle will

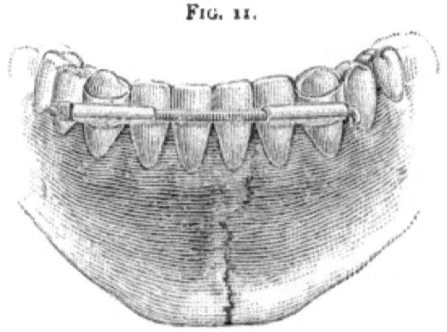

FIG. 11.

enable the reader to comprehend the method peculiar to each class:—

Case No. 1 will illustrate Class No. 1. A young man fell from a pile of lumber, a distance of fifteen or twenty feet, and, besides severe bruises, suffered a simple fracture through the symphysis, terminating, however, in front between the central and lateral on the left side, as shown by the line in the engraving (Fig. 11).

Upon examination, it was found that the fractured bone was quite widely separated at the top, and the left central incisor was loosened. The treatment practiced was as follows: The ends of the fractured bones were placed in their proper position and temporarily fastened by lacing the teeth

INJURIES AND DISEASES OF MAXILLARY BONES. 47

with silk ligatures. Bands of very thin German silver were made to encircle and accurately fit the cuspid teeth. A small tube of German silver, one-half inch in length, was soldered to each band and in exact alignment; a piece of wire accurately fitting the bore of these tubes, bent at right angles at one end and having a screw cut upon the other end, was slipped through each tube and secured therein by adjusting a nut on the screw. The bands were cemented in position upon the teeth by means of oxyphosphate cement, as shown in Fig. 11.

After the cement had become thoroughly set, the nut was then tightened until the fractured ends of the bone were drawn snugly together.

The appliance was worn without displacement or trouble for twenty-one days, when it was removed, the bone having become firmly united. I may add, that during the time the appliance was worn, so firmly was the jaw supported, the patient suffered little if any inconvenience, and after the third day partook regularly of his meals, using his jaws freely, but, of course, avoided the hardest particles of food. After removing the appliance a careful impression of the jaw was taken, a model made, and the appliance transferred to the model, exactly as shown in the engraving. The lower part of the jaw is, of course, diagrammatic, and was added by the engraver to show the line of fracture.

It should be borne in mind that the principle upon which this appliance is based is not the same as where the teeth are simply wired together, but very different; for in wiring, the upper parts of the fracture only are tipped or drawn together, and no pressure or support is given to the lower parts, while in the method here shown it will be seen that, by reason of the bands and pipes being rigidly attached to the anchor teeth, tipping is impossible, and pressure is exerted equally upon both parts (upper and lower) of the fracture as they are drawn together by the screw.

This device may be applied in any locality in either jaw, provided suitable teeth for anchorage be not too far remote from line of fracture. The screw may be bent to accommodate the curve of the arch, should the fracture occur in the region of the cuspid.

These bands, tubes, wires, screws, and nuts are some of the appliances known as "Angle's Regulating and Retaining

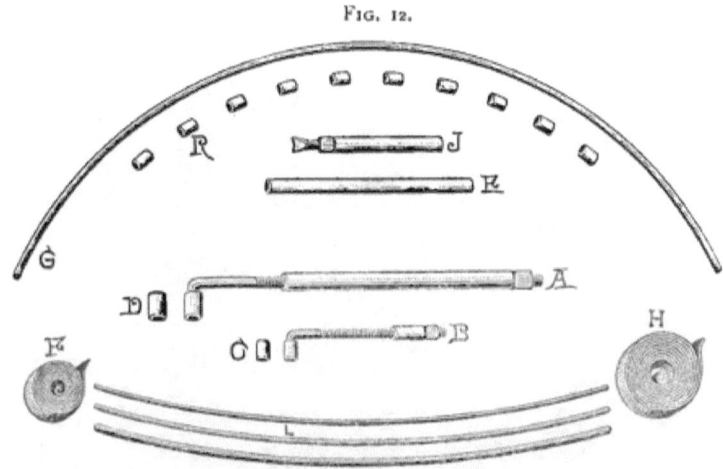

FIG. 12.

Appliances," devised and used for the purpose of correcting irregularities of the teeth. They may be procured of dealers in dental goods. (See Fig. 12.)

The treatment for cases of the second class is illustrated in the following instance: On July 4, 1889, a man aged forty-five was admitted to the Minneapolis City Hospital. A blow from a policeman's club had produced two simple fractures of the inferior maxilla. The first was an oblique fracture of the right side, beginning with the socket of the second bicuspid, extending downward and backward, involving the socket of the first molar, breaking out the second bicuspid, and greatly loosening the first molar. The second molar had been lost years before, while the third, as well as the remaining teeth, were much abraded and loosened by salivary calculus, thus making the application of the appliance described in Case No. 1 impossible. The second fracture was situated on the opposite side high up in the ramus.

Because of swollen condition of the parts, the exact line of fracture could not be detected, but the grinding of the ends

FIG. 13.

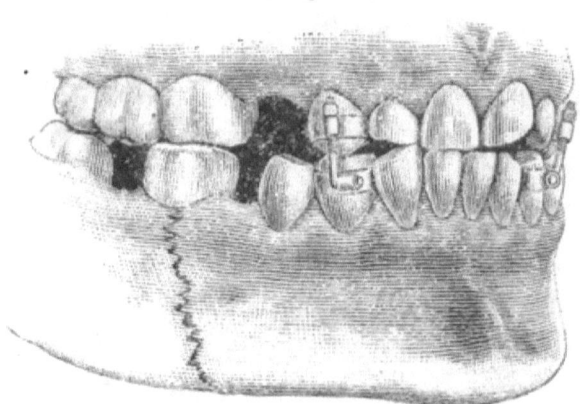

of the bone and the great pain occasioned thereby were unmistakable evidences of a fracture. The patient, as in all such cases, was unable to close the jaws. The fracture on the right side was widely separated, and the anterior piece much depressed by reason of the contraction of the depressor muscles, while the posterior bone was drawn firmly up, the molar teeth occluding. (See Fig. 13.)

The following is the treatment used: Bands were made to encircle all four of the cuspid teeth, they being the most firmly attached in their sockets. The fractured ends of the bones were placed in apposition, the lower jaw closed carefully. The occlusion of the lower teeth upon the upper required so considerable force and occasioned so much pain that it became necessary to anæsthetize the patient. Points on the bands for the necessary attachments were carefully noted. The bands were then slipped off the teeth, and little pipes (shown at Fig. 12) soldered at the necessary points, after which the bands were cemented in their proper position upon the teeth, and two small traction screw-wires, the same as shown at Fig. 12, were slipped into the pipes. The jaws were closed and the nuts tightened on the screws, until the jaws were drawn firmly together, and each tooth occupied its exact position in occluding upon its fellow of the opposite jaw. Both fractures were then carefully examined and found to be in perfect apposition, and presented the appearance shown in Fig. 13. The most natural position for the jaw and the muscles had been secured, thus placing the parts in their natural positions of relaxation and rest.

During an attack of coughing during the night following, one of the bands was wrenched loose, but was replaced the next day without trouble. No further accidents occurred. The patient readily took nourishment through the spaces between the teeth. Thus the fractured jaw was firmly supported without the least motion for twenty-two days, when the appliance was removed, showing most excellent results.

The following case possesses several points of special interest; the fractures were in regions similar to the case just described, and the appliances, though involving similar mechanical principles, will be found to be greatly simplified.

Thomas B. was admitted to the Dental Infirmary of the University of Minnesota, suffering from the effects of a blow received on the left side of the jaw from a canthook, while working in a lumber camp in Wisconsin, which produced fracture of the jaw in two places. The first fracture was on the left side, beginning between the first and second bicuspids and extending downward and backward, and involving the lower part of the anterior root of the first molar. The second fracture was on the right side, directly through the angle of the jaw. The fractures had occurred thirty-two days previous to his admission to the infirmary, during which time nothing had been done to reduce them. He reported that he had called upon a physician, who supposed the trouble was merely an abscessed tooth, and had lanced the gum with the view of reducing the swelling. Later the patient had called upon a dentist in one of the smaller towns, who also failed to diagnosticate the fracture and extracted both bicuspids, in the hope of giving relief. Upon examination I found considerable swelling in the region of this fracture, with the usual result; the patient being unable to close his mouth by reason of the anterior piece of the fractured bone being drawn down by the contraction of the depressor muscles. A false joint had also become established, and the bones could be easily worked without causing pain.

At the point of fracture on the right side there was little or no displacement; the swelling was also slight.

The patient was anæsthetized. The ends of the bone were then rubbed forcibly together with the view of breaking up the false attachments and stimulating activity in repair.

The ends of the bones were now placed in perfect apposition, and the mouth closed, great care being taken to articulate the teeth in their correct position against the upper ones.

The jaw was now firmly bound in this position to the upper teeth, in the same manner as described in case No. 2, with this difference, that the method was improved upon and simplified by using clasp bands, as shown in Fig. 14. No cement was used, and instead of the screws small metallic buttons were soldered to the sides of the bands (as shown in the cut), around which fine binding wire was wrapped in the form of a figure 8. (See Fig. 15.)

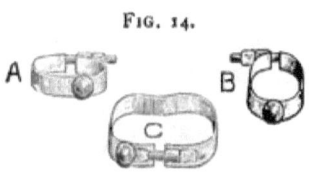

FIG. 14.

The bands seen upon the molar teeth in the engraving were not used in this case, but are shown for the purpose of illustrating how they may be used in case of comminuted fracture.

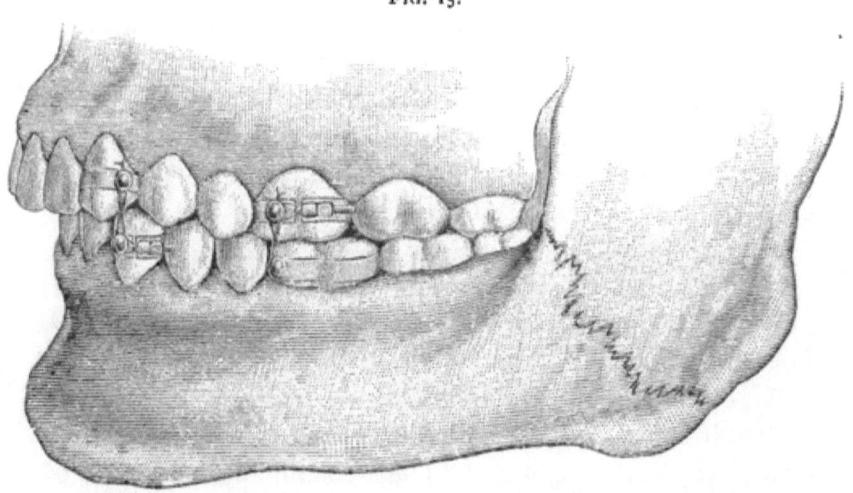

FIG. 15.

At the end of seventeen days the bands were removed and the patient discharged, the bone having been firmly united.

Dr. Angle suggests that it might be urged as an argument

against this method that the teeth being closed and the jaws being firmly bound together the patient would be unable to take sufficient nourishment. It, however, rarely happens that a patient is found without some teeth missing, thereby leaving abundance of space for the passage of liquid foods, and even if all the teeth were sound and in perfect position, it has been proved that there is plenty of space between the teeth and behind the molars and between the upper and lower incisors for taking all nourishment necessary. In such cases more time would be consumed in taking nourishment, but this obstacle is compensated for by the main points of advantage in its favor, such as cleanliness and greater comfort to the patient, as compared with the many bulky and awkward appliances in use.

Thirdly, its extreme simplicity enables any one with ordinary mechanical ability, when provided with a set of clamp bands, to easily and quickly set all ordinary cases of fracture.

And, lastly, the certainty of correct results will be sufficient reason for all those who are interested in this branch of surgery to give it a trial. Class No. 3, comprising fractures of edentulous jaws, are fortunately very rare. The method of treatment is similar in principle to that already described in Class No. 1, only that in place of the teeth small bone hooks are used, drilling for their reception a suitable cavity on each side of the fracture, comparing in position to the original sockets of the teeth, the same as if the operation of implanting teeth were intended ; the cavities thus made need not be nearly so large or deep. They should also be drilled obliquely, to correspond to the course taken by the hooks. The hooks before insertion should, of course, be made antiseptic.

Necrosis of the Jaws.

Necrosis of the jaw is indicated by inflammation, similar to that of dental periostitis. The gum about the affected part is swollen and of a dark red or purple color, pus oozing from the edge of the gum between the teeth, or through one or more fistulous openings; this discharge is not always confined to the mouth; we find the pus escaping sometimes through an opening on the cheek or neck, as low down at times as the clavicle.

The effect of necrosis of the jaw upon the teeth is obvious, since great pain is produced by the least pressure, and in cases of entire necrosis they become loose and discolored; in the majority of such cases the teeth prove such an annoyance to the patient that they are extracted, if they do not drop out of their own accord. Cases are met with, however, where the teeth remain in situation after the bone was both necrosed and had been removed.

Treatment.—Remove the dead portions from around the living bone—here the dental engine and burrs are useful—syringe with tepid water and peroxide of hydrogen to cleanse the parts, followed by stimulants, tonics, and nourishing diet.

New bone is produced in the lower jaw if the periosteum is preserved, this with the surrounding tissue being active in producing new bone to take the place of the lost portion.

In the superior maxillary there is a development of fibrous tissue in the young subject. In the adult nature does not do this much. When a part of the superior maxillary is lost, the periosteum ordinarily makes no effort to repair. (For an elaboration upon this subject see " Heath's Diseases of the Jaw.")

Dislocation of the Lower Jaw.

The causes of dislocation of the lower jaw are yawning, shouting, vomiting, the introduction of the stomach pump, etc. Sometimes it occurs during operations upon the teeth; in all cases the patient's mouth is opened to its fullest extent. The capsular ligament, being very large and tenacious, is not ruptured.

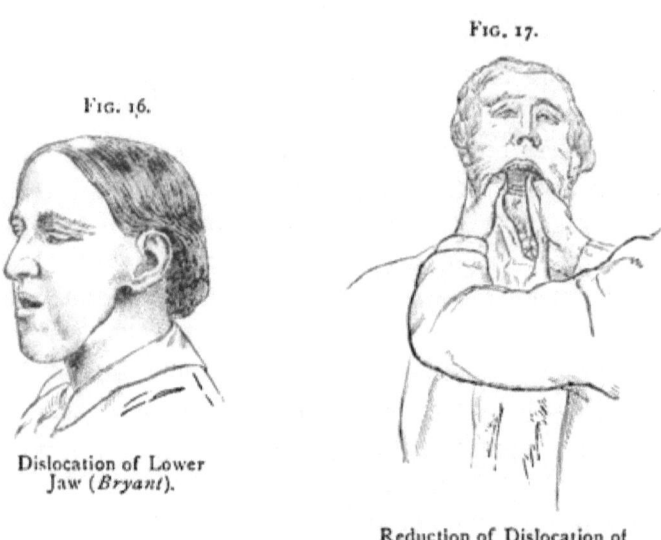

Fig. 16.
Dislocation of Lower Jaw (*Bryant*).

Fig. 17.
Reduction of Dislocation of Lower Jaw (*Bryant*).

The manner in which dislocation takes place is as follows: When the mouth is opened to its fullest extent, each condyle of the jaw leaves the true articular eminence of the interarticular fibro-cartilage, which is drawn forward of the external pterygoid muscle. A cavity is thus left behind the condyle; when the jaw is in this position, but a very slight force is needed to carry the condyle over the articular eminence and produce a dislocation.

Symptoms of Dislocation.—The mouth is open and the jaw fixed, mastication being impossible, the lower teeth project beyond those of the upper. Saliva dribbles from the mouth and speech is indistinct. A careful examination reveals a concavity immediately in front of the ear, and the condyle may be both seen and felt in front of this. The masseter muscle is firmly contracted and very prominent (See Fig. 16).

Treatment.—Reduction is made by placing the thumbs (protected by napkins) as far back upon the molars as possible, depressing the back part of the jaw, followed at once by the raising of the chin, which results in sliding the capitulum backward into the condyle fossa. After correcting the dislocation the jaws should be secured by a bandage extending under the chin and over the top of the head. The patient should be recommended to take care, for some weeks, not to open the mouth too far, as a disposition to a recurrence is great.

INFLAMMATION OF THE TEMPORO-MAXILLARY ARTICULATION.

Serious inflammation of the temporo-maxillary articulation is infrequent; when such complaint arises, it is usually found that some derangement of the teeth is the seat of the trouble; however, when by a close examination we find the teeth and their surrounding tissues in a healthy condition, inflammation of this articulation may be surmised.

Symptoms.—The parts in the vicinity of the joint are sensitive to pressure; they present a swollen appearance and cause considerable pain when the mouth is opened and closed.

Treatment.—First, comparative rest must be given the joint, the patient partaking only of liquid or soft food. An application of some good stimulant, such as tincture of iodine, followed by ice-water compresses, to decrease the heat of the inflamed parts.

ABSCESS OF THE ANTRUM OF HIGHMORE.

The antrum of Highmore is separated from the apices of the superior molars by a very thin lamella of bone (this is sometimes penetrated by the roots of the first molar), and is almost always involved in inflammation in alveolar periostitis. At times this inflammation extends to the mucous membrane of the antrum, but less frequently does an abscess of the alveolar process penetrate this membrane and discharge pus into the antrum. An abscess is sometimes formed in the antrum by the septic decomposition of the pulp of a tooth, which extends into the cavity.

When this does take place, the natural opening between the antrum and the nose becomes diminished by the swelling of the lining mucous membrane. Inflammation becomes excessive, and a large collection of pus takes place which at length escapes into the nose, or burrows alongside of the root of a tooth, and discharges into the mouth. Or a fistula may be established upon the cheek; and in protracted cases the floor of the orbit may give way, and pus discharge along the lower eyelid.

Symptoms.—Abscess of the antrum is indicated by a discharge of pus into the nose, usually accompanied by a bad odor, at times causing severe local and constitutional suffering, or the pus may have other outlets, as given above. The cheek is hot, flushed, and somewhat swollen, pain of a throbbing character is complained of, the malar bone is elevated, with an apparent depression beneath it; the palate in severe cases loses its concavity and becomes convex, and the floor of the orbit is pushed up, forcing the eye partly from its socket.

Treatment.—Make an opening into the antrum, either through the process above the roots of the teeth, or by extracting the affected tooth, and entering by perforating the floor of the antrum (if this is not already done) through the alveolar cavity of one of the roots. The latter procedure is usually practiced. After a free opening is secured, the cavity may be cleansed by syringing it with warm water, followed by peroxide of hydrogen and bichloride of mercury, 1 to 3000.

During treatment a small plate may be worn while eating, to prevent particles of food from entering the cavity. A plate just large enough to cover the cavity, and attached to the adjoining teeth, answers the purpose. Or some temporary stopping, such as cotton and sandarach varnish, or gutta-percha may be placed in the cavity.

DEFECTS OF THE PALATINE ORGANS.

Cleft Palate.—One of the most distressing deformities to which the human frame is liable, is that defective condition of the palatine organs known as Cleft Palate.

It is indicated by a fissure extending through the soft, or both the soft and hard palate, causing an impairment of mastication, deglutition and of speech. They are divided into two classes—Acquired (by accident or disease), and Congenital (dating from birth).

Congenital Cleft Palate is the result of a lack of development of the maxillary bones, which may be caused by hereditary disease, or malformation from lack of nourishment of the tissues involved during embryonic life.*

---

* See article on "Physiology of Voice and Speech," American System of Dentistry, Vol. III.

These defects are sometimes accompanied by more or less deformity of the alevolar arch, and of the teeth, which are usually of a soft texture, with imperfectly developed roots. The cleft is not always confined to the palate bones and the soft palate, but may be complicated with complete fissure of the alveolar process and with harelip.

Acquired Cleft Palate includes all losses of tissue in either soft or hard palate (that are not congenital) whether occasioned by disease or accident. The faculty of distinct articulate speech is impaired, and deglutition is performed with much inconvenience; acquired lesions coming generally in adult life, the individual has not the advantage of the training of the parts during infancy, as in cases of congenital defects.

The infant resorts to a very curious expedient to secure the nourishment necessary for subsistence and growth. The nipple, instead of being taken between the tongue, upper lip, and gum, is taken between the lower surface of the tongue, and the lower lip and gum. This habit being acquired, it is applied later in the mastication of solid food. The food being conveyed between the tongue and movable floor, is brought back between the teeth for deglutition, which is usually performed in this way without any food entering the nose through the Cleft Palate.

Treatment.—The remedy for these deformities must be the closing of the passage in such a way as to restore, as far as possible, to the organs, their functions. This may be done by a surgical operation, or by the insertion of an artificial palate.

STAPHYLORRAPHY.

The Surgical Operation which is sometimes resorted to, is an exceedingly painful one for the patient, and difficult for the operator, and which after all sometimes proves a failure. Indeed, it is claimed by some writers that failures of closure in these operations are in the majority. However this may be, there are a great many cases on record where such operations have been performed with success.

Staphylorraphy is derived from a Greek word, signifying suture of the uvula. It consists in freshening or paring the edges of the palate, and passing ligatures or sutures through, drawing the edges together and closing the gap by tying the sutures.*

It is claimed that the idea of this operation was first conceived by one Le Monnier, an ingenious French dentist, and was successfully performed by him as early as 1764. It does not seem to have been recognized by the medical profession however for more than half a century afterward; it was then, 1820, practiced by M. Roux, of France, and in America by Dr. J. C. Warren, of Boston.

Dr. Warren's methods were considered simpler, and his operations are generally regarded as the basis of the various modifications that have since aided in perfecting the proceeding.

It is now classed among the regular operations of surgery, and I have personally witnessed and assisted in a number of such operations that have been perfectly successful.

Artificial Palates.

Artificial Palates, according to some writers, were employed by the Greek physicians, but the first description was given by the celebrated French surgeon, Ambrose Paré, in the early part of the 16th century. Many improvements upon these primitive forms have been made. They are

---

* For fuller treatment of this operation see "American System of Dentistry," Vol. III.

divided into two distinct classes, obturators and artificial velum.*

An obturator is a non-elastic and stationary cover or stopper for those defects in the hard or soft palate which have a complete and well defined boundary.

An artificial velum† is a movable valve, made to supply the loss of the posterior soft palate; being under the control of the surrounding muscles, opening or closing the passage at will.

EXTRACTION OF TEETH.

The extraction of teeth is an operation that need seldom be resorted to. It is in nearly all cases, from negligence of the patients, or their fear of dental operations, that the teeth are permitted to remain in diseased conditions until they reach such a state that extraction is necessary. It is not surprising that the operation is usually approached with apprehension, since frequent accidents occur in its performance; this, however, is generally due to the neglect, awkwardness or unskillfulness of the operator. It very often occurs in the hands of medical practitioners, and is a subject that should receive more attention by physicians, by whom, though not belonging to their province, it is frequently performed. In fact, if a chair of dental pathology were established in our medical colleges, it would be a wise step in the higher education of medical students.

Indications **Justifying the Operation.**—*First*, with the teeth of first dentition, it is sufficient to state that when a tooth of replacement is about to be erupted, or has actually made its appearance either in front of or behind the corre-

---

* See "American System of Dentistry," Vol. II.

† Velum; a veil, a cover; hence, "a pendulous veil of the palate."

sponding deciduous tooth, the latter should at once be extracted; and when these teeth have been so neglected that they, together with the surrounding tissue, have become seriously diseased, it is best that they should be removed. It is desirable, however, whenever they can be retained in a fair state of health, to do so, retaining the shape of the arch until it is time for their successors to replace them, as well as giving the child their service in mastication.

Second. In regard to the propriety of extracting the permanent teeth it should first be stated that none of these should be sacrificed unless called for by some urgent necessity. Uncontrollable pain and incurable disease surrounding the tooth is an instance.

Third. Extensive loss of surrounding tissue, leaving the tooth or root very much loosened, acting as an irritant and becoming a source of disease to the adjacent parts.

Fourth. Where a tooth is the cause of an incurable alveolar abscess, the offending member should not be allowed to remain; such cases, however, are rare.

Fifth. To prevent or correct irregularity in the arrangement of the teeth.

Sixth. In preparing the mouth for an artificial denture it is sometimes found that the loss of one or more remaining teeth may be advantageous.

There are other cases presented at times, to which fixed rules would not be applicable, where experienced judgment must determine the practice to pursue.

In Conclusion, it is scarcely necessary to say that whenever a tooth can be restored to a healthy condition, it should always be done, and that we should not for any reason be too hasty in extracting the first tooth from an unbroken arch. As all teeth, except the inferior central incisors, have normally two antagonists in articulation, the loss of one tooth would place two others partially without function. And the adjoining teeth would soon become irregular by gradually leaning into the spaces.

Hemorrhage after Extraction.—*In cases where extraction is followed by excessive hemorrhage,* or where the tendency to hemorrhage exists, the application of some reliable styptic should be made, the following being the most powerful of these agents: tannic acid, gallic acid, solution of persulphate of iron, and the powdered subsulphate of iron (Monsel's Powder). Some of the simple local remedies are spider web or a pledget of cotton or sponge saturated with sandarach varnish as mechanical obstructors; this packing should be allowed to remain until all danger of a return is past.

CALCAREOUS DEPOSITS.

There are two varieties of calculus, or tartar, as it is most commonly called, found upon the teeth; namely, salivary and sanguinary.

Salivary calculus is deposited from the saliva upon the crowns or any exposed surfaces of the teeth. It varies in color from a light cream to a dark brown or black, depending upon its age and the habits and general health of the patient. The characteristics of the substance, therefore, furnish diagnoses of importance to the physician and dentist.

Composition.—Salivary calculus is composed of mineral and animal matter; about 75 and 25 per cent. respectively. Phosphate of lime, and in some cases a little magnesia, carbonate of lime, fibrin or cartilage, mucus and a small quantity of animal fat, are its principal ingredients. The relative proportions of its constituents vary according to its density.

All persons are subject to salivary calculus to a greater or less extent, but its physical characteristics are exceedingly variable. Hence it is that analyses made of it by different chemists differ.

Origin and Deposition.—It is generally conceded that this concretion is a deposit, chiefly from the saliva, with an admixture of mucus.

Saliva is the mixed fluid from the different salivary glands and mucous follicles, which in its normal state is either alkaline or neutral. On exposure to the atmosphere, and the decomposition liable to occur in the mouth, it may be found strongly acid or strongly alkaline, holding salts of lime in solution. On the formation of acids in the mouth, and by the action of the carbonic acid exhaled,* decomposition takes place, and the salts of lime are deposited upon the teeth. It is precipitated in great quantities upon the surfaces of the teeth opposite the ducts from which the saliva is emptied, upon the lingual surfaces of the inferior incisors and cuspids, and the buccal surfaces of the superior molars. The necks of the teeth about the free margins of the gum afford favorable points for its collection. A nucleus once being formed, it deposits particle by particle, rapidly encroaching upon the crown, where it is deposited more abundantly. In the mouths of uncleanly persons it sometimes accumulates in such quantities that nearly all the teeth are encrusted.

Salivary calculus is not deposited upon the soft tissues, but upon some substance that forms nuclei, such as the natural and artificial teeth, plates, etc.; though it is found sometimes in the ducts of the salivary glands, owing, no doubt, to a sluggish condition of the saliva.

Treatment.—The removal of salivary calculus is an operation of importance to the health of the gums and alveolar

---

* *Carbonic acid* having a strong affinity for lime, unites with it in the salivary solution, forming carbonate of lime.

process, and the preservation of the teeth. For its removal, instruments (scalers) that may be readily applied to every part of the tooth should be employed. Considerable tact and practice is necessary to perform the operation skillfully. The scalers should be passed well down beneath the margin of the gums, that is below the edge of the deposit, until it is brought in contact with the surface of the tooth, and the mass scaled off in the direction of the cutting or masticating surface. Every particle of the deposit should be removed, care being necessary that the tooth substance is not roughened by the edge of the instruments, and the surface polished, lest a nuclei for immediate reaccumulation be formed.

Chemical agents are employed by some for the removal of salivary calculus. This should be scrupulously avoided, as any acid capable of dissolving this accumulation is more or less injurious to the teeth. "Their careless administration by physicians is a fruitful source of injury to the teeth, and they certainly should form no part of any dentifrice, or be in any way used for the removal of stains of any kind from the teeth." *

Sanguinary calculus † is deposited upon the roots of the teeth, and not upon their crowns, as with salivary calculus. It is precipitated from the liquor sanguinis of the blood, upon its disorganization, in connection with the suppurative process of inflammation. It is in the form of dark granulations, approaching crystallization. It is much harder than salivary calculus, and adheres more firmly.

Composition.—"Sanguinary calculus is composed chiefly of lime salts, colored with the hæmatin of the blood, which increases its tendency to take crystalline form."

---

* Harris's "Principles and Practice of Dentistry."
† See Ingersoll's "Dental Science."

It should be remembered that while salivary calculus causes inflammation, sanguinary calculus is a result of the inflammatory action ; the suppurative process in all cases precedes its deposition.

Mucous Deposit.—The teeth of children are especially liable to mucous deposit or "soft tartar." It is generally found upon the labial surfaces of the superior front teeth, though it is occasionally found upon the same surface of the inferior teeth.

Source.—It can readily be seen that it is not precipitated by the saliva, from its not collecting upon the surfaces of the teeth opposite the mouths of any of the ducts leading from the salivary glands. It is generally considered to be deposited from the mucus. It is most abundant when the mucus is secreted in large quantities and when this secretion is in an abnormally acid condition.

Color.—It varies in color, from a grayish-brown to a dark green.

The Effect of Mucous Deposit upon the Teeth.—"This green stain so erodes the enamel that decay advances in the part which it covers more or less rapidly, according to the quality of the teeth and the length of time it is allowed to remain. The removal of this mucous deposit requires more skillful manipulation than that of salivary calculus, on account of its being a thin film entering into the substance of the enamel, rendering it difficult to detach without injury to the tooth substance." *

Treatment.—Where the erosion is but slight, this mucous deposit may be removed by finely powdered pumice stone and water, applied on a piece of some hard, fine-grained wood, such as orange wood, or on wood points, rotated by means of

---

* Harris's "Principles and Practice."

the dental engine, the surface being left thoroughly polished or burnished.

When the effects are more serious, the enamel not only being discolored, but deeply eroded, the use of the corundum point, rotated by the dental engine, is necessary ; after which the surface should be left well polished.

DENTAL MEDICINE.

Dental materia medica is an embodiment of the nature, medicinal properties, and therapeutical action, of all substances used as medicine in dental practice.

The classification of medicines is made according to their action upon the animal economy.

The different classes in common use by dentists are as follows: Narcotics and hypnotics, analgesics or anodynes and anæsthetics, stimulants, tonics, sedatives, antipyretics, irritants, astringents, styptics, and hæmostatics, caustics, escharotics, antizymotics or antiseptics, and disinfectants and laxatives.

NARCOTICS AND HYPNOTICS.

Narcotics (stupor) are medicinal substances which, by impairing or destroying nervous action, lessen the relationship of the individual to the external world. They at first, however, have a stimulating effect, to which their therapeutic efficacy is largely due, which is followed by profound sleep and stupor. If the dose be sufficient, death will ensue by paralysis of the centres of the medulla, which govern respiration and the other functions of organic life.

Hypnotics (sleep) belong to the class of narcotics, but are capable of causing sleep without any preliminary cerebral excitement, by bringing the brain into a favorable condition for it.

The principal narcotics are opium (see Anodynes), alcohol (see Stimulants), belladonna (see Anodynes), chloroform, ether (see Anæsthetics), etc.

The hypnotics are opium, the bromides, chloral, etc. When administered to relieve pain, they are termed anodynes.

BROMINE, Br.—BROMIDES.

Derivation.—Bromine is obtained from sea-water and certain saline springs.

Properties.—Bromine is a dark brownish-red, liquid, non-metallic element. It has an offensive, suffocating odor, somewhat resembling chlorine and iodine. In its pure state it is an active escharotic and internally a violent poison. The salts of bromine are cerebral and cardiac depressants and are highly valued as hypnotics.

The Principal Preparations:—

Ammonium Bromide, NH_4Br.—Colorless, prismatic crystal. Dose, gr. v–xx.

Calcium Bromide, $CaBr_2$.—A white, granular, deliquescent salt. Dose, gr. v–ʒj.

Lithium Bromide, $LiBr$.—A white, granular, deliquescent salt. Dose, gr. v–xx.

Potassium Bromide, KBr.—Colorless, cubical crystals. Dose, gr. v–ʒj.

Sodium **Bromide**, $NaBr$.—Colorless, monoclinic crystals. Dose gr. v–ʒj.

Zinc Bromide, $ZnBr_2$.—A white, granular, deliquescent powder. Dose, gr. ss–ij.

Syrup of Bromide of Iron.—Contains 10 per cent of ferrous bromide, $FeBr_2$. Dose, ʒss–j.

Therapeutic Uses.—The bromides are used as sedatives

to the nervous system to produce sleep, and in affections of the heart or cerebrum, when shown by increased action, in neuralgia, spasmodic cough, etc.

Dental Uses.—Bromide of potassium is a useful remedy in convulsions from the irritation of dentition, in neuralgia, also in cases of extreme sensitiveness of the soft palate. Dose, gr. 10-20 every hour for several hours before taking impression.

CHLORAL, C_2HCl_3O.

Derivation.—Chloral is obtained by the action of chlorine gas on absolute alcohol. It is a colorless, unstable, oily fluid, which readily combines with water and forms *chloral hydrate*, the official "chloral" having the formula $C_2HCl_3OH_2O$.

Properties and Actions.—The official body, chloral hydrate, is in the form of a white, crystalline substance, having a pungent odor and taste, and is soluble in water, alcohol, and glycerine.

It is hypnotic, antispasmodic, and to a limited degree anæsthetic. It is serviceable in fevers, accompanied by cerebral excitement, convulsions, delirium tremens, etc. Dose, from v to xxx grs. Liebreich claims to have produced profound sleep, lasting from five to fifteen hours, with twenty-five to thirty grains.

The hypnotic action is preceded by a stage of excitement of short duration, which is followed by sudden calm and refreshing sleep, from which the patient can be easily aroused to partake of nourishment and will readily fall asleep again—differing in this respect from narcotism, which is marked by profound stupor.

Dental Use.—Hydrate of chloral is sometimes used in dental practice for the relief of odontalgia from pulpitis, from one-half to one grain being applied to the inflamed body. It has also been thought a serviceable agent by some in the treatment of putrescent pulp-canals, and as a stimulant and antiseptic injection in chronic alveolar abscesses.

ANALGESICS OR ANODYNES.

Anodynes are agents which are capable of relieving pain. They are divided into two classes, general and local.

General **anodynes,** when taken internally, affect the whole organism, by depressing the cerebral centres of perception and sensation.

Local anodynes, when applied, affect the parts either by impairing the conductivity of the sensory nerve fibres, or by reducing the local circulation. Some of the most efficient anodynes act either general or local. The principal agents of this class are as follows:—

General Anodynes.—Opium, morphia, belladonna, aconite, ether, and chloroform (see Anæsthetics).

Local Anodynes.—Opium, belladonna, carbolic acid (see Escharotics), cocaine (see Anæsthetics), aconite, etc.

OPIUM.

Source.—Opium is obtained from the white poppy, an annual herb grown in Asia Minor.

Nature.—It is a gummy exudation which follows the incising of the unripe capsules. It should yield not less than nine per cent. of morphine when in its normal moist condition.

Opium contains seventeen alkaloids, the most important of these being *morphine*—dose, gr. $\frac{1}{20}$–$\frac{1}{2}$—hypnotic, narcotic, and anodyne.

Principal Preparations of Opium:—

Pulvis opii, powdered opium. Dose, gr. $\frac{1}{6}$–ij.

Tinctura opii (laudanum), composed of powdered opium, oz. iiss; and diluted alcohol, Oj (pint). Dose, ♏xij, or 25 drops, equivalent to 1 gr. of opium.

Tincture opii camphorata (camphorated tincture of opium, paregoric) is prepared by macerating "sixty grains of opium in two pints of diluted alcohol, with sixty grains of benzoic acid, a fluidrachm of oil of anise, two ounces of clarified honey, and forty grains of camphor." Dose, f ʒj-f ʒj. Dose for infant, v to xx drops (gtt.)—ʒss. contains about gr. j. It therefore contains $\frac{1}{20}$ the strength of the tincture.

Pulvis ipecacuanhæ et opii (Dover's powder), composed of ipecac 1 part, opium 1 part, sugar of milk 8 parts, triturated to a fine powder. Dose, v to xv grs.

BELLADONNA.
(*Deadly Nightshade*.)

Source and Composition.—It is an European plant, the leaves and root being the medicinal portions. It contains two alkaloids—atropine, the active principle, and belladonnine.

Preparations of Belladonna. From the leaves :—

Tincture of Belladonna.—Dose, ♏j-ʒss.

Extract of Belladonna.—Dose, gr. ¼-½.

From the root :—

Abstract of Belladonna (powdered).—Dose, gr. ⅛-½.

Fluid Extract of Belladonna.—Dose, ♏j-v.

Sulphate of Atropine.—Dose, gr. $\frac{1}{100}$-$\frac{1}{60}$.

Therapeutics.—Belladonna is especially useful in the pain of inflammation, particularly that of rheumatism, neuralgia, etc., and is used locally in connection with morphine to relieve the pain of abscesses, boils, etc.

Atropine is used by ophthalmologists to lessen pain, dilate the pupils, paralyze the accommodation, etc.

ACONITE.

Source and Composition.—It is obtained from the tuberous root of *Aconitum napellus*, a perennial plant, found in the mountainous regions of Europe and Asia. The leaves are sometimes used, but the root makes the most powerful drug.

The active **principle** is the alkaloid aconitine, a sedative poison.

Principal preparations:—

Extract of Aconite.—Dose, gr. ½-j.

Fluid Extract of Aconite.—Dose, ♏ ¼-ij.

Tincture Aconite.—Dose, ♏ ss.-iv.

Medical Properties and Action.—Aconite is a powerful sedative to the nervous system. In large doses it acts as a cardiac, respiratory, and spinal depressant.

It proves fatal in poisonous doses by paralyzing the heart and respiration. It is also diaphoretic and antipyretic.

Dental **Therapeutics.**—Aconite, in the form of a tincture, is administered in inflammatory affections and in chronic cases of neuralgia. It is an active antagonizer of the fever process, and has been termed the "therapeutic lancet."

When applied locally, it checks inflammation in its first stages, by paralyzing the peripheral ends of the nerves in the parts, and favoring resolution; also limits the extent of an abscess where pus has already formed.

In combination with the tincture of iodine, in equal parts, it acts very promptly in the incipient stages of dental periostitis, relieving the inflammation, retarding the circulation, and stimulating lymphatic action.

In such cases the gum over the affected tooth should be thoroughly dried and then painted with this combination, protecting the lip or cheek until the remedy is absorbed. It is also considered useful by many in the dressing of pulp canals, preventing the formation of inflammatory products. When applied to a large surface, or where the skin is abraded, care should be exercised, or dangerous constitutional effects may result.

The physiological antagonists are, atropine, morphine, digitalis, and ammonia. *In aconite poisoning* the stomach should be evacuated, stimulants administered, warmth applied to the extremities, and the recumbent position maintained.

ANÆSTHETICS.

Anæsthetics are agents which temporarily destroy sensation and relieve pain. They are generally employed for this purpose during surgical operations. They are divided into general and local anæsthetics.

General anæsthetics are volatile substances, capable of producing (when inhaled) complete unconsciousness, loss of sensibility, and lessened motor power.

The principal agents of this class are, ether, chloroform, nitrous oxide gas, and bromide of ethyl.

Local anæsthetics are agents whose action is limited to the circumscribed parts to which they are applied. They paralyze the nerves of the part, thus temporarily destroying sensation. They act similarly to the local anodynes, except that while the anodynes diminish the sensibility of the parts, the local anæsthetic destroys sensation entirely for a time.

The principal agents of this class are, cocaine, absolute ether, aconite, atropine, etc.

ÆTHER—ETHER, $C_4H_{10}O$.

Derivation.—Sulphuric ether (improperly so called) is ethylic ether or oxide of ethyl. It is obtained by the distillation of ethylic alcohol and sulphuric acid, the acid dehydrating the alcohol and remaining in the retort. $(C_2H_6O)_2 - H_2O = C_4H_{10}O$.

Medical Properties and Actions.—Ether is a colorless, volatile, and inflammable liquid. It is an anæsthetic and anodyne, a diffusible stimulant, and a narcotic poison. Administered internally, it is one of the most powerful secretion stimulants known. The action of the heart, and hence the circulation, is increased, flushing and warmth of the surface soon follow. The senses are more keen, and the phenomena of alcoholic intoxication results, which is less protracted, however, ether being quickly eliminated, chiefly by the lungs.

Principal Preparations:—

Æther Fortior, stronger ether, ethyl oxide, "$C_4H_{10}O$," contains about six per cent. of alcohol.

Sulphuric Ether, ethyl sulphate, $C_4H_{10}SO_4$.

Nitrous Ether, ethyl nitrite (sweet spirit of nitre), $C_2H_5NO_2$. The well-known antipyretic and diaphoretic. Dose, ℥ v-ʒ ij.

Dental Use.—Ether is employed as a general and local anæsthetic, as a local anodyne in neuralgia and odontalgia, and as a counter-irritant, evaporation being prevented.

Ether as an Anæsthetic Agent.—The practicability of producing anæsthesia by the inhalation of ether was first demonstrated by Dr. Horace Wells, of Hartford, Conn., and Dr. W. G. S. Morton, of Boston, Mass., during the years 1844-'46.

Ether, though less prompt in its action, is much safer than chloroform. It has its necrology, however; a number of fatal cases (about thirty) have been reported.

The Administration of Ether.—The operator should be well assured, before administering an anæsthetic, that the patient is not laboring under any serious disease of the heart, brain, or lungs, as ignorance in this direction might lead to fatal results. The clothing about the neck and chest should always be loose, lest it act as an impediment to respiration, and if artificial teeth be worn, they should be removed before the administration of the anæsthetic.

For the inhalation of ether and chloroform a number of instruments have been devised, but the simplest and probably the best method is from a sponge, napkin, or handkerchief, placed within a cone, formed of a towel or stiff paper, with a small opening at the apex for the admission of air, or a small piece of lint can be held in the palm of the hand and on these pour the anæsthetic agent.

The inhalation should be commenced cautiously, the patient should be directed to breathe quite naturally, and to obey any instructions given, as the raising of the hand, etc. The towel or napkin should be held three or four inches from the patient's face, approaching it gradually, thus overcoming the irritating effect and a sense of strangulation, which follow when the agent is placed at once to the mouth and nostrils.

Action of Ether.—The first stage of anæsthesia is a slight relaxation, the second is tetanic or convulsive, the third, complete relaxation.

During complete anæsthesia the face is cool, there being a profuse perspiration; the eyes are closed, insensible to the touch, and the pupils are somewhat contracted. The respiration and pulse are somewhat slower than normal, as shown in the following table:—

Normal pulse, 72 per minute.

Pulse on administration of ether:—

1st min.	2d min.	3d min.	4th min.	5th min.
92	109	110	94	69

Normal respiratory movements average about 20 per minute.

Respiration on administration of ether :—

1st min.	2d min.	3d min.	4th min.	5th min.
23	24	26	18	15

Order in which Nerve Centres are Acted Upon.— First, the cerebrum ; second, cerebellum ; third, the spinal cord ; fourth, the medulla oblongata.

The Quantity of Ether Required.—Largest quantity, 9 ozs. ; minimum, 2½ ozs. ; average quantity to produce anæsthesia, 5 ozs.

The Time Required for Full Anæsthesia.—Longest time required, 24 minutes ; shortest time, 3½ minutes ; average time, 8 minutes.

The Dangers of Anæsthesia.—There are conditions rendering general anæsthesia dangerous, and the practitioner, whether medical or dental, should be well assured, before administering ether or chloroform, that none of these are present. They are fatty degeneration of the heart, valvular lesions, kidney disease, brain tumors, respiratory obstructions from enlarged tonsils, thoracic tumors or aneurism, and chronic alcoholism. An anæsthetic should never be administered on a full stomach, as sickness would likely follow that would interfere with the operation, and anæsthesia of the glottis prevents the expulsion of vomited matter in case it enters the larynx by regurgitation ; neither should it be given after long fasting, as an absence of nutrition would tend toward cardiac paralysis ; excitement should be avoided, instruments should be kept out of sight, and too many spectators should not be present. A painful operation should not be commenced before the stage of complete anæsthesia is reached, or it may cause death from shock, as the result of peripheral irritation.

Treatment of Dangerous Symptoms.—In case of the suspension of the heart's action, the agent should be withdrawn, the body placed in a reclining position, and, if needs be, inverted, and air freely admitted. The failure of respiration requires the drawing forward of the tongue, by a finger being thrust deeply into the mouth; the inhalation of a good stimulant, nitrite of amyl, gtt. ij to gtt. v; but care is necessary in its use and not more than two or three drops should be administered to patients who have never inhaled it. The inhalation of ammonia is probably as efficient, and can be used with more freedom than nitrite of amyl. Galvanism, too, has been successfully employed as a cardiac and respiratory stimulant, "the positive pole being placed to the nostril and the negative pole over the diaphragm, to excite a reflex action between the fifth pair and the pneumogastric, or the poles may be placed directly over both phrenic nerves, or on a line with the fourth cervical vertebra, in order to stimulate respirations; or one pole may be placed over the upper dorsal spinous process and the other pole over the apex of the heart, to induce cardiac contraction."* And if necessary, artificial respiration should be employed (see page 135) and warmth applied. The extremities should also be rubbed briskly, rubbing upward.

Note.—It should be remembered that ether vapor is heavier than air, and forms therewith a highly explosive mixture. Therefore, if a light must be in the room, it should be high above the patient. A grate fire, gas stove, etc., in the vicinity are very dangerous.

CHLOROFORM, $CHCl_3$.

Derivation.—Chloroform (Ter-chloride of Methyl, or Methylic Ether) is obtained by distilling alcohol with chlorinated lime. It was discovered in 1831 by Samuel Guthrie, of Sackett's Harbor, N. Y.

---

* Gorgas's "Dental Medicine."

The form for medicinal use is Chloroformum Purificatum, or Purified Chloroform.

Medicinal Properties and Action.—When inhaled, chloroform is an anæsthetic, and when administered internally, it is an anodyne and antispasmodic. If swallowed undiluted, it excites great inflammation of the mucous membrane and causes violent gastritis.

Its effects are similar to those of ether, but more rapidly produced, and it is more powerful in its action; hence, requiring more care in its administration.

When first administered (internally), it causes a feeling of warmth in the stomach, which is soon followed by a sense of coldness. It increases the action of the heart, producing excitement of the brain, followed by depression and deep, heavy sleep. In large doses it causes stupor and insensibility, and has caused death.

Therapeutic Uses.—Chloroform is used for the same purposes as is ether, and is much employed locally in liniments. Administered by inhalation, it is a general anæsthetic, and when administered internally, in substance, it is an anodyne and antispasmodic, and is used as such in cases of nausea, sea-sickness, sick headache, and in cases of cholera. In the last named it has probably proven more efficacious than any other single remedy. Dose, ℳj–ʒss, diluted, internally.

Spirit of chloroform (chloroform, ʒj; diluted alcohol, ʒij), ʒss–ʒj.

For inhalation, ʒj–ʒj. Average, ʒiij.

Dental Use.—Chloroform is employed by some in dental practice as a general anæsthetic; its use, however, is growing less every year, in favor of ether and nitrous oxide gas. It is also used as a local anæsthetic; in this case it is generally combined with other substances, as aconite, alcohol, ether, opium, etc.

For administration as an anæsthetic, treatment of dangerous symptoms, etc., see Ether.

Chloroform Narcosis.—

> Shortest time, ... 2 minutes 30 seconds.
> Longest time, ... 14   "   30   "
> Average time, ... 6   "   24   "

Chloroform mortality is 1 in 3000 (over 500 fatal cases are reported, none of which were in obstetrical practice).

Compared with ether mortality, 1 in 16,000.

NITROUS OXIDE GAS, N_2O.

History.—Nitrous Oxide, or "Laughing" Gas, was discovered by Dr. Priestley in 1776, and its respirability demonstrated by Sir Humphry Davy, though the results were not published until some twenty years afterward. In 1844, Dr. Horace Wells practically demonstrated the value of its anæsthetic property for the relief of pain during surgical or dental operations.

Nitrous oxide gas is manufactured by slowly melting and boiling the salt nitrate of ammonia in a glass retort, dissolving it into a vapor of water and a permanent gas ($NH_4NO_3 +$ Heat $= N_2O + 2H_2O$). The gas should pass through three wash bottles, the first containing a solution of the sulphate of iron or caustic potash, and the other two pure water, for the purpose of purifying it before it enters the receiver, from which it is administered to the patient through an inhaling tube. A pound of the salt will generate about thirty gallons of the gas. It is perfectly fused at 226° F., white fumes are emitted at 302° F., and gas begins to evolve at 460 F. If the temperature is raised to 500° F., a dangerous impurity, nitric oxide, is given off; this need not be generated, however, if the proper care is observed, not allowing the temperature to rise above 480° F.

Liquefied Nitrous Oxide.—The most convenient form for use is the liquefied gas, it being liquefied and solidified under intense cold and great pressure (50 atmospheres, or 750 pounds pressure). It is then secured in strong iron cylinders, from which it is allowed to escape into an inhaling bag when needed for use.

The advantages of this form of gas are its purity, convenience for use, the large supply which can be kept on hand, and its comparative freedom from deterioration, notwithstanding its age.

Properties and Actions.—Nitrous oxide gas is an elastic, colorless gas, with a very slight and agreeable odor. It will freeze into a beautiful, clear, crystalline solid, at about 15° F. below zero.

"By the evaporation of this solid, a degree of cold may be produced far below that of carbonic acid bath *in vacuo*, or lower than 17° F."*

Nitrous oxide gas supports combustion with nearly the same promptness as oxygen.

As an Anæsthetic.—Nitrous oxide gas is the most pleasant and the safest general anæsthetic known. The shortness of the anæsthetic stage is the greatest objection to its administration for surgical operations, though its rapid action, comparative safety, and the transient nature of its effects on the system render it the most useful anæsthetic agent for all minor operations, such as the extracting of teeth, removal of nerves from the teeth, where the tooth substance is lost to such

---

* Gorgas's "Dental Medicine."

an extent that a devitalizing agent could not be retained, for the lancing of abscesses, etc.

The Administration of Nitrous Oxide Gas, for dental operations, should be conducted with the same care that is given to ether and chloroform, though it is a comparatively safe anæsthetic. The patient should be seated in an operating chair which will admit of the back being lowered to such a degree that the patient could at once be placed in a horizontal position. The dress about the throat and waist, if tight, should be previously loosened, and the patient should not have partaken of food for at least two hours previous to the inhalation of the gas.

A mouth prop, of which there are several patterns manufactured, *should be placed between the teeth,* to prevent the closure of the jaws, as the muscles become rigidly contracted during the administration of this gas. The most suitable prop is one made of India rubber—the ordinary lead-pencil eraser, cut in proper lengths, answers the purpose very nicely—or a firm cork, as it also prevents injury to the teeth or fillings, as sometimes occurs when a mouth prop of some hard substance is used. The patient is then directed to take full, regular, and deep inspirations of the gas, the nose being held or covered, to prevent the admixture of atmospheric air. Its anæsthetic effects are soon made manifest by strong, involuntary respirations, accompanied by snoring, this being caused by the relaxation of the muscles of the pharynx, and paralysis of the tongue, causing it to fall back toward the throat, interfering with breathing, and a livid appearance of the lips, cheeks, and finger nails, which is due to the discolored blood in the capillaries. But the most delicate test for complete anæsthesia is, as in ether and chloroform, the loss of sensibility to the touch in the conjunctiva of the eye.

The amount of gas required to produce complete anæsthesia varies, from five to fifteen gallons being the usual amount. Out of 2000 administrations I have had one case where 65 gallons were required, and another where 80 gallons were inhaled before the anæsthetic stage was reached.

The **first stage** under nitrous oxide gas is muscular activity

The second **stage** is muscular rigidity. It cannot be continued until complete muscular relaxation, lest the patient die of asphyxia.

Nitrous **Oxide Gas Mortality.**—There is about one death to each 125,000 administrations.

Dangerous **Symptoms, with Treatment,** etc.—See Ether.

BROMIDE OF ETHYL, C_2H_5Br.

Derivation.—Bromide of ethyl, or hydrobromic ether, is obtained by distilling bromide of potassium and sulphuric ether, and redistilling with chloride of lime.

Properties.—It is a colorless, volatile fluid, possessing an agreeable ethereal odor and a pungent taste. It is not inflammable, caustic, nor irritant; in this respect it is preferable to chloroform or ether as an anæsthetic agent.

The Administration.—Bromide of ethyl is administered as is ether or chloroform, or in a folded starched napkin, so as to cover the face, as directed by Prof. Gorgas. A soft linen handkerchief is placed inside the napkin, and upon this the agent is poured; one drachm should be used at first, directing the patient to take deep, full inspirations. At the end of two minutes the second drachm should be added; this should be repeated at intervals of two minutes, until complete anæsthesia is produced. The quantity differs according to the susceptibility of the patient.

Action.—The administration of bromide of ethyl is attended with some danger, and clinical experience has not demonstrated to careful operators that it is as safe as some other and older agents of this class. It has a toxic action on the centres of respiration. The heart force is decreased and its action is more frequent, which contributes to the paralysis of the respiratory centres. Several deaths occurred in a very limited number of administrations of this agent.

COCAINE.

Source.—Cocaine is the active crystalline alkaloid of Erythroxylon coca, a small Peruvian shrub. The leaves resemble those of Chinese tea, and in South America they are used by eight millions of people, much as we use tea or coffee.

In the preparation of the alkaloid, it is necessary that the leaves be carefully gathered, as the best quality only should be used. They should be dried, and not injured by age or exposure to the air, as moisture deprives them of value.

Preparations of Erythroxylon :—

Extractum Erythroxyli Fluidum, fluid extract of erythroxylon. Dose, ℨ ss–ij.

Salts of cocaine.

Cocaine Hydrochlorate, $C_{17}H_{21}NO_4$.—Dose internally, gr. ⅛–ij; most commonly used as a local anæsthetic in aqueous solutions, 2–5 per cent.

Cocaine Oleate, cocaine and oleic acid, 5–20 per cent. solutions for external use.

Cocaine Hydrobromate, cocaine and hydrobromic acid, 2–10 per cent. as a local anæsthetic.

Cocaine Wines, Pastes, Lozenges, etc., are made in great varieties.

Medical Properties and Action.—Cocaine, when applied locally, acts as an anæsthetic; when taken internally in small doses, it is a general stimulant, improving digestion, stimulating the respiration, circulation, etc. It produces wakefulness and a marked diminution of the sense of fatigue and hunger. For this reason the leaves are chewed by the Peruvian Indians to sustain them during long journeys or arduous labor.

A toxic dose, or long-continued use (cocaine habit) produces insomnia, decay of the moral and intellectual powers, hallucinations, insanity, and death.

Dental **Use.**—The salts of cocaine have proven very efficient for their local anæsthetic and anodyne effects; their power as a local anæsthetic is very great over a limited area, and hence it is of special value to the dentist, for operations upon the submucous tissues and the extraction of teeth, where it should be used by hypodermic injection, or applied to the gum on either side of the tooth to be extracted, the latter method being the safer; two or three applications should be made at intervals of about two minutes each, when a painless operation is generally secured. I have found it to act very happily also, in connection with arsenious acid, for the devitalization of dental pulps, the pulp dying without giving the patient any discomfort. But as for its use as a pain obtundent in hypersensitive dentine, its practical benefits are questionable.

A warning, however, should be given, that a potency for evil lurks in this most valuable drug. In many cases where it has been injected into the gum tissue for extraction of teeth, toxic results of an alarming nature have occurred and patients have been rendered ill for several weeks. This, however, is not apt to follow when the patient is of a sanguine temperament and in good health. I have made a record of many cases where toxic results have followed the use of this drug, and find them all to be of a nervous or hysterical temperament, or pregnant women. The lesson is that we should use judgment and discrimination in its application.

Dangerous Symptoms.—The extremities usually become cold and rigid, the eyes staring and glassy, and the face pallid, while the pulse is weak, the heart beats faint, and respiration slow and weak—the symptoms of an impending collapse.

Treatment of Dangerous Symptoms.—Fresh air should be admitted and some good stimulant administered, such as brandy, or aromatic ammonia and nitrite of amyl—by inhalation, or ether in case of convulsions, and if need be the battery. As soon as the patient is able, assist him to stand up and promenade.

CHLORIDE OF ETHYL, C_2H_5Cl.

Properties.—Chloride of ethyl is a colorless liquid possessing a strong ethereal odor, and is very volatile and inflammable in ordinary temperature. Its boiling point is about 50° F. It is due to this low boiling or vaporizing point that it is so exactly adapted to the special requirements of a local anæsthetic. It is put up in convenient glass tubes, drawn out to a fine point, and hermetically sealed.

The point of the tube is marked by a file scratch at its smallest part. Here the point is broken off when ready for use, either by the fingers or the forceps. Immediately the chloride in a gaseous state escapes from the small opening, and if the tube is partially inverted, a small jet of the liquid is projected; this is further accelerated by allowing a good portion of the tube to come in contact with the hand, the warmth of which hastens the vaporization of the liquid.

The preparation of chloride of ethyl, as spoken of above (in glass tubes), is a patented process, controlled by P. Monnet, of Lyons, France. As the chloride evaporates in ordinary temperature and is very inflammable, the point has to be drawn out and sealed while the tube and its contents are immersed in ice water.

The Application and Action.—When about to apply, the parts to be anæsthetized should be thoroughly dried, by means of absorbent cotton or napkin, then the point of the tube should be broken as previously directed, and the fine jet of chloride directed upon the surface.

If teeth are to be extracted, a napkin should be placed in the mouth back of the teeth to be operated upon, and the patient directed to breathe entirely through the nose; the liquid should then be projected upon the mucous membrane around the tooth or root and upon the cheek over the track of the inferior maxillary nerve for the lower, and on the temple over the emergence of the fifth nerve for the upper teeth. This application upon the face, however, need not be made unless the teeth are very difficult to extract, and prolonged anæsthesia is desired.

It is seldom necessary to use the entire contents of a tube for a single operation; from one-quarter to a half will usually produce complete anæsthesia of the parts. The opening in the tube can then be closed and the contents preserved for a subsequent operation by folding a small piece of rubber dam and placing it over the end, and over this stretch a wide rubber band, from end to end; then place the tube in an upright position, preferably in a glass of cold water.

The writer has employed chloride of ethyl in over two hundred minor surgical operations with uniform success. It is a most satisfactory local anæsthetic in the extraction of teeth, lancing of abscesses, removal of small tumors, extraction of the tooth pulp, and in the preparation of roots of teeth and the fitting of bands and caps in crown and bridge work.

STIMULANTS.

Stimulants are medicinal agents which increase organic activity. The most powerful and rapid in action, though transient in effect, are termed diffusible stimulants, while the local stimulants, which are of a vegetable nature, containing a volatile oil, are termed aromatic.

Among the first class are such agents as the alcoholic preparations, ammonia, camphor, ether, nitrite of amyl, myrrh, etc.

The principal members of the class of aromatic stimulants are capsicum, oil of cloves, peppermint, etc. Heat and cold also act as local stimulants.

ALCOHOL, C_2H_6O.

Derivation.—Alcohol is obtained by repeated distillations from the product of fermented grain or starchy substances, easily converted into grape sugar, which in the presence of and by the growth of low vegetable organisms (the yeast plant, etc.) splits up into alcohol and CO_2. Commercial alcohol contains about 90 per cent. of absolute alcohol with 10 per cent. of water.

Properties and Action.—Alcohol is a colorless, inflammable fluid, wholly vaporizable by heat, and unites in any proportion with water and ether. It possesses a pungent odor and burning taste. All of the alcoholic preparations are powerful diffusible stimulants, causing general exhilaration of spirits.

In large doses, however, it is a depressant, producing muscular incoördination and the effects of narcotic poisons, ending in delirium, coma, and death.

The Most Important Alcohols are—

Methylic Alcohol, C_2H_4O, methyl hydrate, wood spirit.

STIMULANTS.

Ethylic Alcohol, C_2H_6O, ethyl hydrate, grain spirit.

Amylic Alcohol, $C_5H_{12}O$, amyl hydrate, potato spirit, also occurs with the ethylic alcohol, in excessive distillations of fermented grain.

Principal **Preparations of Alcohol** :—

Absolute **Alcohol,** rarely obtainable in the shops, however, stronger than 98 per cent.

Alcohol contains about 91 per cent. of absolute alcohol.

Alcoholis **Dilutum**, contains equal parts of alcohol and water.

Spiritus **Frumenti**, whisky from rye, corn, barley, and potatoes, contains from 45 to 50 per cent. of alcohol.

Rum, obtained by the distillation of fermented molasses.

Wines—port wine, sherry white wine (made by fermenting the juice of the grape without the seeds, stems, or skins); red wine (from the juice of grapes with their skins); champagne, claret, Rhine, etc. These contain from 5 to 40 per cent. of alcohol.

Beer by slow fermentation, contains		2 to 3 per cent. alcohol.		
Ale, by rapid "		"	2 to 6	" "
Porter and Stout		"	4 to 6	" "

Therapeutic Uses.—The alcoholic preparations are most valuable agents in disease, for appropriate cases; they are employed as stimulants in acute inflammations, such as pneumonia, pleurisy, bronchitis, phthisis, and in the last stages of typhoid fever, diphtheria, acute neuralgia, etc. In insomnia, from cerebral anæmia, small doses of some alcoholic stimulant at bedtime are found beneficial. In poisoning by cardiac depressants and snake venom, alcohol, freely, sustains the heart. In chloroform anæsthesia, an ounce of whisky beforehand will sustain the heart and prolong narcosis.

Dental Uses.—In the administration of nitrous oxide gas, a small quantity of wine taken beforehand will often be found beneficial, increasing the heart's action at about the time the effects of the anæsthetic are passing off. In painful operations upon the teeth, I have found small doses of sherry or champagne to be very beneficial.

As a styptic, it arrests hemorrhage by coagulating the blood by its effects upon albumin, and contracts the mouth of the vessels by its astringent properties.

For suppurating wounds it is a useful antiseptic dressing.

For the treatment of softened and sensitive dentine, and for drying cavities preparatory to filling, absolute alcohol is generally an efficient agent; after drying the cavity with cotton or bibulous paper, it should be bathed with alcohol, which evaporates rapidly and causes the almost perfect absorption of moisture from the dentine.

Treatment of Acute Alcoholism.—Evacuate the stomach, administer ammonia cautiously by inhalation, apply warmth to the extremities and cold affusion to the head, and, if needs be, artificial respiration.

AMMONIA.

Medicinal Properties and Action.—It exists most commonly in the form of ammonia gas, NH_3; which, dissolved in water, is the aqua ammonia of commerce. It is intensely alkaline, and is an irritant to the mucous membrane. When inhaled it acts as a stimulant, especially as an antagonist of cardiac depressants. Prolonged inhalation induces spasmodic coughing, a sense of suffocation, and inflammation and œdema of the glottis; when swallowed, the aqua sets up violent inflammation of the passages and stomach.

The salts of ammonia, in medicinal doses, are stimulating expectorants, and stimulate the heart's action; while in large doses or continued use they produce rapid emaciation, by impairing digestion and increasing tissue waste. In large doses they also injure the red blood corpuscles.

Principal Preparations:—

Aqua ammonia, water of ammonia, containing 10 per cent. of the gas in water. Dose, ♏v–xxx, diluted.

Aqua ammoniæ **fortior,** containing 28 per cent. of the gas in solution.

Ammonium carbonate. Dose, gr. ij–x.

Ammonia chloride, sal ammoniac. Dose, gr. j–xx.

Ammonia spirits (a 10 per cent. solution of aqua ammonia in alcohol). Dose, ♏x–℥j, diluted.

Aromatic **spirits of ammonia,** the carbonate with aromatics (oil of lemon, lavender, etc., and alcohol and water). Dose, ♏x–℥ij.

Ammonia liniment, aqua ammonia, 30 per cent., and cotton-seed oil, 70 per cent.

Ammonium **nitrate,** used in preparing nitrous oxide gas.

Ammonium **sulphate,** used in preparing other ammonium salts, etc.

CAMPHOR, $C_{10}H_{16}O$.

Source.—Camphor is a white, concrete, and translucent gum, obtained from the volatile oil of the camphor laurel, an evergreen tree indigenous in China, Japan, Formosa, etc. Refined camphor is prepared in large circular cakes, one to two inches thick.

Properties and Action.—Camphor is slightly soluble in water (about 1 to 1300), but freely in alcohol, ether, chloroform, oils, and milk. Alcohol forms a 75 per cent. solution. It has a penetrating, fragrant odor, a bitter, pungent taste, leaving a slight sense of coolness. It is a stimulant, anodyne, diaphoretic, antiseptic, and irritant.

In medicinal doses it temporarily increases the heart's action, stimulates respiration and mental activity, promotes perspiration, and allays pain and spasm.

Large doses depress the heart and excite narcotic symptoms, and have proved fatal.

Principal Preparations :—

Aqua camphora, camphor water (8 parts of camphor to 1000 of distilled water, with 16 parts of alcohol to aid in the suspension of camphor). Dose, ℨ j–iv.

Spiritus camphoræ, spirit of camphor (camphor, ℨ iv, alcohol, Oj). Dose, ♏ v–xx.

Linimentum camphoræ, camphor liniment (camphor 1 part to olive oil 4 parts).

Linimentum saponis, soap liniment (soap 10 parts, camphor 5, oil rosemary 1, alcohol 70, water 15). Is an anodyne and mild irritant for sprains, rheumatic pains, etc.

Dental Use.—In dental practice the spirit of camphor is sometimes employed by local application to allay the pain of sensitive dentine, and that which sometimes follows the extraction of teeth, and the wounding of pulps of teeth. Camphor is also employed in the treatment of putrescent root canals of teeth. It is also one of the ingredients of the celluloid base for artificial teeth.

NITRITE OF AMYL, $C_5H_{11}NO_2$.

Derivation.—Nitrite of amyl is produced by the action of nitric or nitrous acid upon amylic alcohol.

Properties and Action.—Nitrite of amyl is a clear, yellowish, oily liquid. It has an ethereal odor, and is very volatile and inflammable; it is insoluble in water, but soluble in alcohol, ether, and chloroform. It is used by inhalation, causing great cardiac activity, vascular dilatation, flushing of the face, a sense of fullness of the brain, and complete resoution of the muscular system. It is a muscle poison, and when the vapor is applied directly to the muscular or nervous tissues it arrests their functional activity.

Dental Use.—Nitrite of amyl, being a powerful stimulant to the heart, is employed in syncope and chloroform narcosis. It is also used in epileptic attacks and other convulsive or spasmodic diseases. Cases are reported where nitrite of amyl has restored the patient after artificial respiration had failed. Care, however, must be observed in its use, as it is a powerful and dangerous agent.

Dose of nitrite of amyl by inhalation is from ♏ ij to ♏ v. Not more than two or three drops should be administered to weak and nervous patients who are susceptible to its influence.

MYRRH.

Source.—Myrrh is a resinous exudation from a small tree grown in Arabia and the northeastern coast of Africa, known as the *Balsamodendron myrrha*. It is a spontaneous exudation from the stems of the tree, which collects in small masses upon the bark.

Properties and Action.—Myrrh is brittle and is easily pulverized. It is of a reddish-yellow color, translucent, with an aromatic taste and a peculiar fragrant odor. When pulverized the powder is of a light yellow color, if pure. In medicinal doses myrrh is a stimulant and astringent. It stimulates the digestive organs and improves the appetite, but in larger doses it acts as an irritant to the gastro-intestinal membrane. It is employed externally as a local application to inflamed, ulcerated, and relaxed tissues, for its stimulating and astringent effect.

Dose.—Powdered myrrh, gr. x to ℥ ss, in pill form or suspended in water. **Tincture of myrrh,** ʒ ss to j (myrrh ℥ iij, alcohol O ij.)

Dental Use.—The tincture of myrrh, diluted, forms an excellent gargle and mouth-wash, and a stimulating lotion for spongy and inflamed gums. The powder is employed as an ingredient of many dentifrices, for its astringent properties.

CAPSICUM.

Source.—Capsicum, or Cayenne Pepper, is the fruit of *Capsicum fastigiatum*, a plant of tropical Africa and America. Its pungent odor and hot taste are due to its very acrid and volatile principle, called capsicine.

Medicinal Properties and Actions.—Capsicum in medicinal doses is a powerful stimulant. It produces a sensation of warmth in the stomach, and a general glow over the body; it stimulates the circulation and digestive process, but in excessive doses it acts as an irritant poison.

Preparations:—

Tincture Capsicum. Dose ♏ v–ʒ j.

Powdered Capsicum. Dose gr. v–x in pills.

Emplastrum Capsicum. A most excellent plaster.

Dental Use.—Capsicum in tincture or plaster form, preferably the latter, is very serviceable in dental periostitis, as it aids in establishing resolution or hastens suppuration. It is also an excellent stimulating gargle, tinct. capsicum ʒ ss to rose water ʒ viij.

OIL OF CLOVES.

Source.—The oil of cloves is obtained from the dried, unexpanded flowers of the *Eugenia caryophyllata*, an evergreen tree of the myrtle order, a native of the Indies.

Properties and Actions.—The oil of cloves, when fresh, is a clear and colorless preparation; it has a pungent, spicy taste, and a fragrant odor. Is an aromatic stimulant, irritant and antiseptic. It is sometimes administered to relieve nausea, and prevent griping when combined with purgatives, also to modify the action of other medicines. Dose, ♏ j-v.

Dental Use.—The oil of cloves is employed in dental practice to relieve odontalgia, by introducing two or three drops into the carious cavity of the aching tooth, relieving the pain by its stimulating effect upon the pulp. It is sometimes used for the same purpose in combination with other agents, and has the effect of rendering carbolic acid more pleasant, without interfering with its action. It is used also by microscopists to clarify preparations for mounting.

Eugenol ($C_{10}H_{12}O_2$) is an active principle of oil of cloves. It is sometimes called an acid, as it possesses some acid qualities. It is a clear, colorless oil, and its odor and taste resembles that of the oil of cloves. It is an excellent antiseptic for dental uses.

PEPPERMINT.

Source.—Mentha piperita, or peppermint, is grown everywhere, and as a plant is familiar to every one. The leaves and tops are used for medicinal purposes.

Properties and Actions.—The properties of peppermint are due to a volatile oil, in which form it is generally used. It is an aromatic stimulant, carminative and antispasmodic, and local anodyne and anæsthetic when evaporation is prevented after being applied to the surface.

Preparations :—

Oil of Peppermint (consisting largely of menthol). Dose, ♏ j-v.

Peppermint Water (2 parts of the oil to 1000 of distilled water). Dose indefinite.

Essence of Peppermint (10 per cent. of oil with 1 per cent. of the powdered herb in alcohol). Dose ♏ x-xxx.

Dental Use.—Local anodyne and anæsthetic.

TONICS.

Tonics are agents which give healthful activity and vigor to the functions, gradually imparting strength and tone to the system, that is, without preternatural excitement. They are divided into vegetable and mineral tonics.

Principal among the vegetable tonics are cinchona, nux vomica, digitalis, cimicifuga, and eucalyptus.

While the principal mineral tonics are the preparations of iron, arsenic, zinc, sulphuric acid, nitric acid, muriatic acid, etc.

CINCHONA (PERUVIAN BARK).

Source.—Cinchona is the bark of any variety of cinchona. The different species of this tree are natives of the mountains of western South America, especially in Peru and Bolivia, though they have been planted and are grown in India, Ceylon and Burmah.

The medicinal properties of these barks depend upon the alkaloids they contain, which are in varying proportions, usually from 3 to 4 per cent., at least 2 per cent. of which is quinine, this being the most important.

Principal preparations of cinchona and its alkaloids are as follows:—

Powdered Cinchona. Dose, gr. x–ʒ iij.

Tincture Cinchona (strength, 20 per cent). Dose, ʒ j–℥ ss.

Extract Cinchona (in pill). Dose, gr. j–x.

Sulphate of Quinine. Dose, gr. j–xx.

Sulphate of Cinchonidine (one-half the strength of quinine). Dose, gr. ij–xxx. Much used in hospital and dispensary work.

Properties and Actions.—The different varieties of cinchona are named according to their color. Yellow cinchona —cinchona flava; pale cinchona—cinchona pallida; red cinchona—cinchona rubra. The powder from the yellow bark is of an orange color; has a more bitter taste than the other barks, containing more of the alkaloid quinine. Cinchona is a bitter tonic, astringent, antipyretic, and antiseptic. The alkaloid quinine is preferable for ordinary use, as a much larger quantity of the powdered bark is necessary to obtain the full effects, often causing derangement of the stomach, headache, and constipation.

Dental Uses.—In dental practice quinine is employed in from five to ten grain doses as a tonic, and in the treatment of neuralgia when due to malaria. Cinchona is also used as an antiseptic. "The powder dusted over unhealthy wounds will arrest putrefaction, and promote healthy cicatrization. Quinine will destroy minute organisms, and preserve substances from decomposition."* Cinchona is also employed for its antiseptic and tonic properties as an ingredient in certain dentifrices.

NUX VOMICA.

Source.—Nux vomica is the seeds of the *Strychnos nucis vomica*, a tree of the family Strychnoides, which grows in India. These seeds have been long sold in the shops under the names of nux vomica, bachelor's buttons, poison nuts, etc., and for a long time were used only for such purposes as poisoning rats.

Medical Properties and Actions.—Nux vomica contains two alkaloids, strychnine and brucine, to which its medicinal properties are chiefly due. Brucine has only $\frac{1}{12}$ the strength of strychnine, but they are otherwise identical, physiologically and therapeutically.

---

* Gorgas's "Dental Medicine."

In small doses nux vomica is a bitter tonic, exciting the secretions and stimulating the functions of the body.

In full doses (strychnine gr. $\frac{1}{10}$) the function of the spinal cord is exalted, causing tetanic spasms of the extensor muscle, the lower jaw is stiff, the pupils dilated, and the face wears an unmeaning smile.

In toxic doses (strychnine gr. $\frac{1}{2}$) the function of the spinal cord is paralyzed, respiration is arrested, death following from asphyxia; consciousness is preserved, however, until CO_2 narcosis takes place,

Treatment of Strychnine Poisoning.—The antidote is tannic acid, which forms an insoluble tannate; then the stomach pump should be employed or emetics administered, after which the patient should be kept perfectly quiet.

The antagonists are chloral, chloroform, and potassium bromide; the last named, though, is rarely used, on account of its being so slow of action.

The bladder must be evacuated frequently, lest a re-absorption of the poison take place.

Preparations:—

Abstract of Nux Vomica. Dose, gr. $\frac{1}{2}$ gradually increased to gr. j.

Tincture of Nux Vomica (20 per cent of the drug). Dose, ṁj–x.

Extract of Nux Vomica. Dose, gr. $\frac{1}{8}$–j.

Fluid extract of Nux Vomica. Dose, ṁj–v.

Sulphate of Strychnine. Dose, gr. $\frac{1}{100}-\frac{1}{20}$.

Dental Uses.—Where a cardiac or nerve tonic is required, nux vomica and its chief alkaloid hold the first rank.

DIGITALIS.

Source.—Digitalis, or foxglove, is the leaves of *Digitalis purpurea*, or purple foxglove: the leaves of the second year's growth are considered the best. The plant grows wild in Europe, and is cultivated in this country, where it is sometimes seen in private gardens, grown for its beautiful spike of purple flowers. The Shakers cultivate it quite extensively for the drug market.

Medical Properties and Actions.—Digitalis is chiefly used in disease for its tonic and diuretic properties, its tonic effect upon the heart, principally; though the heart is slowed by its action, its force is at the same time increased. For the full cardiac effects the recumbent posture should be maintained. *When the doses are large*, severe gastric disturbance is caused. *In toxic doses*, the muscles and peripheral nerves are paralyzed; respiration is first slowed and then becomes rapid and feeble; coma and convulsions followed by death from the sudden paralysis of the heart.

Preparations and Doses.—

Digitalis (the leaves). Dose, gr. ss–iij.

Abstract of Digitalis. Dose, gr. ¼–j. (strength, 200 per cent.

Extract of Digitalis. Dose, gr. ⅙–j.

Fluid Extract of Digitalis. Dose, ♏j–iij.

Tincture of Digitalis (15 per cent.). Dose, ♏v–xx.

CIMICIFUGA.

Source.—Cimicifuga, or the black snake root, is the root of the *Cimicifuga racemosa*, a common plant in the United States.

Medical Properties and Actions.—Cimicifuga has a bitter and nauseous taste, somewhat resembling that of opium. It is an efficient cardiac tonic, antispasmodic, diaphoretic, and diuretic. It is feebler in its action than digitalis, and should be used more frequently when the latter drug is indicated.

Preparations and Doses :—

Fluid Extract of Cimicifuga.   Dose, ♏ v–xxx.

Tincture of Cimicifuga (20 per cent. in strength). Dose, ♏ xx–lx (ʒ j).

EUCALYPTUS.

Source.—Eucalyptus is obtained from the leaves of the *Eucalyptus globulus,* or "Blue Gum Tree," a native of Australia, but is now grown in Northern Africa, Southern Europe, and in the United States.

Properties and Actions.—The leaves are the only portion of the tree which possesses medicinal qualities, the fresh being more active than the dried leaves. Their medicinal properties are due to a volatile oil, called oleum eucalypti, which contains three oils, eucalyptene, turpene, and cymol, which distill over at different temperatures, the first product being the most important. Eucalyptus promotes appetite and digestion, and increases the heart's action.

In large doses it causes indigestion, nausea, diarrhœa, and great muscular weakness, and if continued will cause irritation and congestion of the kidneys. It is eliminated by the skin, bronchial mucous membrane, and kidneys, the secretions of which become strongly odorous, owing to the presence of the oil. Eucalyptus is also an antiseptic, disinfectant, sedative, and diaphoretic, "and has anti-malarial properties, absorbing noxious germs, as well as enormous quantities of water from the soil, and by its emanations purifying the atmosphere in the vicinity. It is largely cultivated in malarial districts for these properties, and has rendered habitable a portion of the deadly Roman Campagna."*

---

* Potter's "Materia Medica."

Preparations of Eucalyptus :—
Extract. Dose, gr. j–xv.
Fluid Extract. Dose, ♏xx–ʒj.
Tincture. Dose, fʒ ss–ij.
Oil. Dose, ♏v–xx in emulsion or capsules.

Dental Use.—In dental practice the oil of eucalyptus is employed either alone or combined with iodoform, for its antiseptic properties, in the treatment of putrescent pulps of teeth and chronic alveolar abscesses. This combination has also proven very efficient in the treatment of necrosis and caries of the bone of the jaws.

By taking advantage of the solvent effect of eucalyptol upon the gutta percha, it will be found of great benefit in the insertion of fillings of this material.

IRON.

Ferrum or *iron* is a metal of a bluish-gray color, fibrous in texture, is hard, ductile, malleable, and magnetic. Chemical analysis demonstrates the presence of iron in the blood, 1 part to 230 of red corpuscles, also in the gastric juice, chyle, bile, lymph, urine, milk, and pigment of the eye.

Properties and Actions.—Iron taken into the stomach in the metallic state, meeting with the acids of that cavity, is dissolved; which causes an evolution of hydrogen gas, and gives to the iron molecular activity. *Given medicinally in small doses*, the salts of iron act through and upon the blood, improving its quality and increasing the number of red corpuscles; they also promote the appetite and improve digestion, and hence it is recognized as one of the most efficient tonics.

In large doses, these salts cause nausea and vomiting and act as irritants. Or, the prolonged administration of small doses exhaust the gastric glands by over stimulation.

Monsel's preparations of iron are principally used externally, for hemorrhage, and are considered to be among the very best styptics in use. When internally employed it is for their hemostatic effect in hemorrhage from remote organs. In administering iron care should be exercised, as nearly all the preparations are more or less astringent, and act injuriously on the teeth.

Contra-indications.—Iron should never be given when plethora (a superabundance of blood) exists, especially when accompanied with a hemorrhagic tendency.

Principal Preparations :—

Tincture of the Chloride of Iron. Dose, ♏ v–xx.

Powdered Sulphate of Iron, Monsel's Powder. Dose, gr. ss–iij, in pill; used also as a styptic.

Solution of Subsulphate of Iron, or Persulphate, Monsel's Solution. Possesses powerful astringent properties; used only as a styptic.

Dental Uses.—In dental practice the chief indications for iron is where hemorrhage follows the extraction of teeth, or from any other cause, such as wounds of the gums and mucous membrane. Monsel's solution or powder is employed for this purpose. *See chapter on Extraction of Teeth.*

ARSENIC, As.

Properties.—Arsenic is a brittle, granular metal, of steel-gray color, is very combustible, and volatilizes before melting, the vapor having an odor like that of garlic. It is a powerful poison, not of itself, however, but by virtue of the facility with which it absorbs oxygen. It is generally found in cobalt ore. It is not employed as medicine in its native state.

Preparations :—

Arsenious Acid (?), White Arsenic, "Ratsbane." Dose, gr. $\frac{1}{60}$ to $\frac{1}{10}$.

Solution of Arsenious Acid, 1 per cent. solution (strength, $\frac{1}{100}$) with hydrochloric acid and distilled water. Dose, ♏ ij–x, after meals.

Solution of Potassium Arsenite, Fowler's solution (strength, $\frac{1}{100}$). Dose, ♏ ij–x, after meals.

White Oxide of Arsenic (As_2O_3), Arsenious Acid, is in the form of irregular solid lumps, having a chalky appearance externally, though it is often perfectly transparent internally. It is usually furnished in the shops, however, in the form of a fine white powder, and is often adulterated with chalk or lime. It is odorless and has a faint sweetish taste.

Physiological Actions.—*In small doses*, arsenic is a general tonic, promoting the appetite, digestion, and cardiac action, stimulates mental activity, and causes rotundity of form and clear skin. *In large doses* it becomes a violent corrosive poison, creates skin eruptions and itching of the eyelids, nausea, dysentery, and an irritable and feeble heart, death following from narcotism. *Externally*, it is a powerful escharotic.

Toxicology.—The antidote to arsenic is the hydrated oxide of iron. After the prompt evacuation of the stomach, this should be administered, the dose being eight times the quantity of the poison taken. This should be followed by mucilaginous or oily drinks, to protect the mucous membrane, and iodide of potassium or alkaline mineral waters, to promote elimination.

Tests for Arsenic.—There are a number of tests for arsenic, the following being considered the best: If in a solid state, place the suspected material on burning charcoal, when the arsenic, if present, will become deoxidized, and emit the garlic odor spoken of above. When in an aqueous solution, it may be detected by adding sulphide of ammonium, which produces a yellow sulphide of arsenic, or the addition first of ammonia, then a small quantity of nitrate of silver, will produce a light yellow arsenite of silver. Again, the addition of potassa and sulphate of copper produces a light green arsenite of copper.

Marsh's Test.—The most delicate test for arsenic consists in subjecting the material to the action of nascent hydrogen.* The arsenic is deoxidized and forms with the hydrogen arseniuretted hydrogen gas; this also has the peculiar odor of garlic, burning with a bluish-white flame, which deposits metallic arsenic in the form of a black spot on the surface of a cold plate if held directly in the flame.

Reinsch's test consists of boiling the material suspected of containing arsenic with hydrochloric acid and copper foil, when, if arsenic is present, it will manifest itself in the form of a coating of gray metallic arsenic upon the foil.

Dental Uses.—Arsenic is employed in dental practice for its devitalizing power in destroying the vitality of the pulps of teeth. It is generally combined with other agents, in the form of paste or fibre, for this purpose. But I have found the white arsenic alone to act very happily when applied to the pulp and retained by a small pledget of cotton which had been *previously saturated with cocaine.* The cavity should be completely secured, that none of the arsenic come in contact with the part outside of the tooth.

The **quantity** to be used for this purpose is about the $\frac{1}{25}$ of a grain, and the time required is usually about 24 hours, though there are instances where 48 hours or more are required to thoroughly destroy the vitality.

---

*Nascent hydrogen is evolved by the action of diluted sulphuric acid on zinc.

ZINC, Zn.

Properties.—Zinc is one of the metallic elements—it is very hard, has a bluish-white color, and the fresh surface has considerable lustre, but is soon dulled, from the facility with which it oxidizes.

Principal **Preparations**:—

Zinci Oxidum.   Dose, gr. j–x, insoluble in water.

Zinci Acetas.   Dose, gr. ¼–ij ; as a lotion, gr. ij to ℨj of water, in which it is very soluble.

Zinci **Sulphas.**   Dose as a tonic and astringent, gr. $\frac{1}{10}$–j. As an emetic, gr. vj in ℨiv of water, in tablespoonful doses, repeated every few minutes until emesis takes place.

Zinci Carbonas Præcipitat.   As ointment, or dusted over wounds as a protection.

Zinci Iodidi.   Dose, gr. ss–v, in the form of a syrup.

Zinci Chloridum, tonic and escharotic.   Dose, gr. ss–ij, well diluted.

Zinci Chloridum Liquor, solution of chloride of zinc, ½ to 1 per cent. in strength.

Physiological **Actions.**—The salts of zinc are more or less poisonous, the soluble salts, the acetate, sulphide, and chloride being corrosive poisons. *In small doses* they are tonic and astringent, while in larger quantities they are strong emetics.

The sulphate is a specific emetic, acting without much depression.

The chloride is a powerful and penetrating escharotic. It is also a useful deodorizer and disinfectant. "When applied to malignant and indolent ulcers, it promotes healthy granulations, and when topically applied it not only destroys the diseased structure, but excites a new and healthy action of the surrounding parts."

The antidotes for zinc poisoning are, the white of an egg, carbonate of soda, magnesia, etc.

Dental Uses.—In dental practice the *chloride of zinc* ($ZnCl_2$) is a valuable agent. It is employed as an obtunding agent for sensitive dentine—the sensitive surface being previously bathed with chloroform, which will modify the painful action of the chloride. It has also been employed as a styptic to arrest superficial hemorrhage from a wound of the gum during the filling of the teeth. It induces union of the wounded parts by first intention, by its effect upon the glutinous matter, also as an injection for chronic alveolar abscess, and in diseases of the antrum of Highmore. It is also used in the recession of the gum and the absorption of the alveolar process from the necks of the teeth. The application can be conveniently made by means of a piece of orange wood, so shaped as to permit of its being introduced beneath the gums.

The chloride of zinc, in solution, is also used as one of the ingredients of the filling material known as the oxychloride of zinc, the other ingredient being the oxide of zinc.

The combination of these two forms of zinc makes an excellent capping material, and is probably the best of all materials for root filling.

The oxide of zinc—ZnO—is sometimes employed, combined with carbolic acid, in the form of a paste, for capping exposed pulps; it is also one of the ingredients of the zinc filling materials, and of the celluloid base of artificial teeth.

The sulphate of zinc—$ZnSO_4 7H_2O$—is sometimes employed in disease of the antrum of Highmore, and ulcerations of the mucous membrane, for its stimulant and astringent properties.

SULPHURIC ACID, H_2SO_4.

Properties.—Sulphuric acid, or oil of vitriol, is a dense, inodorous, colorless, oily, and corrosive liquid. It consists of not less than 96 per cent. sulphuric anhydride and about 10 per cent of water.

Preparations :—

Sulphuric Acid. Used as an escharotic or caustic.

Diluted Sulphuric Acid (10 per cent. of the acid to 90 per cent. of water). Dose, ♏ v–xv, well diluted.

Aromatic Sulphuric Acid (Elixir of Vitriol). Sulphuric acid diluted with alcohol and flavored with ginger and cinnamon (strength 20 per cent.). Dose, ♏ v–xxv, well diluted.

Action.—The action of sulphuric acid in its different forms is as follows: *Aromatic sulphuric acid*, tonic and astringent; *diluted sulphuric acid*, tonic, astringent, and refrigerant (in fevers); *sulphuric acid*, escharotic.

Treatment of Sulphuric Acid Poisoning.—Being a corrosive poison, sulphuric acid causes death from asphyxia (the suspension of vital phenomena, from the non-oxygenation of the blood—an excess of carbon dioxide). Administer alkalies, as washing soda, magnesia, lime water, soapsuds, etc., to neutralize the acid, and mucilaginous drinks freely, to protect the mucous membrane. Stimulants, opium, ammonia intravenously, to combat the depressed condition of the vital powers.

Dental Uses.—The concentrated sulphuric acid is employed in dental practice as a caustic; in the laboratory, in a diluted state, for the cleansing of metals before and after soldering ("the acid bath"). It is also used in the manufacture of pyroxylin—gun cotton.

Aromatic sulphuric acid is more agreeable for use in the mouth, while its action resembles that of diluted sulphuric acid. It is a valuable agent in the treatment of pyorrhœa alveolaris and necrosis of the maxillary bones, stimulating the parts to healthy action. It is also employed in the treatment of chronic alveolar abscesses, in combination with a few drops of tincture of capsicum.

NITRIC ACID, HNO_3.

Properties.—Nitric acid, or aqua fortis, is a highly caustic liquid, very volatile, its fumes being corrosive and suffocating, and in the pure state is colorless and transparent, but that usually found in shops is of a yellow color, owing to the presence of nitric peroxide. *Strong nitric acid* is never given internally; it is used in the form of the *diluted nitric acid*, 10 per cent. absolute acid. Dose, ♏ iij-x, well diluted.

Action.—Pure nitric acid is a powerful caustic and escharotic, and is rarely used except as an application to foul, indolent ulcers, or to warts. The diluted acid is a tonic, alterative, and refrigerant, used as a drink in fevers. It is, as are most mineral acids, injurious to the teeth; hence, care should be taken in its use. It should be taken through a glass tube or quill, and followed by an alkaline mouth wash. It is not as agreeable to the stomach as diluted sulphuric acid.

The antidotes for nitric acid poisoning are magnesia or soap and mucilaginous drinks.

Dental Uses.—Nitric acid is employed in dental practice as a caustic for malignant ulcers of the mouth, and has been used for devitalizing pulps of teeth when nearly exposed by mechanical abrasion. It is also used in combination with hydrochloric acid (aqua regia) as a solvent for gold.

PHOSPHORIC ACID, H_3PO_4.

Properties.—Phosphoric acid is a solid, colorless compound, soluble in water and vitrifiable by heat (converted into glass). It is obtained from bones, where it exists in combination with lime. *Diluted phosphoric acid* is the form in which phosphoric acid is usually employed in medicine. It contains 10 per cent. of the absolute acid. Dose, ♏ v-xx.

Action.—Phosphoric acid is tonic and refrigerant, and in large doses an irritant poison. It has been employed externally in the treatment of osseous tumors and caries of the bones.

Glacial phosphoric acid, $HOPO_5$, is obtained from calcined bones. They are first treated with sulphuric acid, "which produces an insoluble superphosphate of lime, then dissolving out the latter salt and saturating it with carbonate of ammonia, which generates phosphate of ammonia in solution, and, finally, obtaining the phosphate of ammonia by evaporating to dryness, and then igniting it in a platinum crucible. The ammonia and all of the water, except one equivalent for each equivalent of the acid, are driven off, and the glacial phosphoric acid remains. It is a white, transparent, fusible solid, generally in the form of sticks, inodorous and sour to the taste. It slowly deliquesces, and is sparingly soluble in water, but freely soluble in alcohol." *

Dental Uses.—Phosphoric acid has been employed in dental practice as a local treatment of osseous tumors and caries of the maxillary bones.

Glacial phosphoric acid is employed as one of the ingredients of the plastic filling material, known as oxyphosphate of zinc, the other ingredient being the white oxide of zinc.

HYDROCHLORIC ACID, HCl.

Properties.—Hydrochloric or muriatic acid is nearly colorless when pure, but that usually found in the shops is of a pale yellow color, being contaminated with chlorine, iron,

---

* Gorgas's "Dental Medicine."

and other substances. It is volatile, emitting a dense white and suffocating vapor; taste very acid and caustic.

Actions.—Hydrochloric acid is caustic, escharotic, and disinfectant. The diluted acid administered internally is tonic, refrigerant, and astringent.

Diluted Hydrochloric Acid (ten per cent. solution of absolute acid and water). Dose, ♏ v–xx.

Dental Uses.—It is sometimes a useful application for treatment of ulceration and inflammation of the mucous membrane and gums. "The strong acid is employed in the laboratory for dissolving zinc, in the preparation of a flux for soldering certain metals."

SEDATIVES.

Sedatives are agents which exert a soothing influence, that is, diminish pain, by lessening the functional activity of organs.

The principal agents of this class are opium and aconite (see Anodynes), digitalis (see Tonics), alcohol (see Stimulants), chloroform (see Anæsthetics), etc.

ANTIPYRETICS.

Antipyretics are agents which reduce the temperature of fever. They act either by lessening heat production or by radiation of heat.

The most prominent of this class are antipyrine, quinine (see Tonics), aconite, alcohol (by increasing heat radiation), also cold bath, ice to the body, etc.

ANTIPYRINE, $C_{20}H_{18}N_4O_2$.

Derivation.—Antipyrine is an alkaloidal product of the destructive distillation of coal-tar oil.

Properties.—It is a whitish, crystalline powder, soluble in water (one-half its weight of hot and its own weight of cold water); less soluble in alcohol, chloroform, and ether; is slightly bitter and odorless. It may be administered hypodermically; is non-irritant to the stomach or the tissues. When combined with ferric chloride it gives a bright red color, and with nitric acid a beautiful green color.

Actions.—Antipyrine is a powerful and popular antipyretic, a general anodyne, hæmostatic, and also possesses mild anæsthetic and hypnotic powers. A full medicinal dose (gr. xxx) produces a stimulant stage of short duration, which is soon followed by profuse sweating, coolness of the surface, slowed pulse, and more or less depression. The temperature in fevers is reduced from 2 to 10 degrees in from 1 to 5 hours, according to the size and continuance of the dose. In health the reduction of the temperature is very slight, and it gives rise to slight nausea and depression. It is eliminated by the kidneys, appearing in the urine a few hours after taking.

"In toxic doses its principal influence is exerted upon the blood, altering the shape of the red blood corpuscles, separating the hæmatin, and causing decomposition of that fluid."*

Dose, for adult, gr. v–xxx; children, gr. j–x.

Dental Uses.—Antipyrine may be employed in dental practice for its hæmostatic, anæsthetic, and anodyne powers.

IRRITANTS.

Irritants are agents which produce more or less vascular excitement or inflammation. They may be either chemical, mechanical, or nervous.

Chemical irritants being those which act by virtue of

---

* Potter's "Materia Medica."

their affinity for organic tissue, exciting the action of the capillaries, and causing an afflux of vascular and nervous power to the part to which they are applied. Included in this class are iodine, capsicum (see Stimulants), turpentine, ammonia (see Stimulants), etc.

Mechanical irritants are agents or means that cause lesions or inflammation by mechanical operation. Filling material or other foreign substance being forced through the apical foramen of the root of a tooth will cause sufficient irritation to produce an abscess; and cuts, contusions, etc., are included in this class.

Nervous irritants act through the medium of the nerves, as nervous shock, depression, or sympathetic inflammation.

IODINE.

Derivation.—Iodine is obtained principally from marine plants, though it occurs in cod-liver oil and shell-fish to a limited extend.

Properties.—Iodine is a non-metallic element, is usually in the form of bluish-black crystalline plates or scales, having a metallic lustre, a peculiar odor, hot, acrid taste, and is of neutral reaction. It volatilizes at a low temperature, giving off a beautiful purple vapor, is slightly soluble in water (1 in 7000), readily soluble in alcohol and ether (1 in 12), also in a solution of chloride of sodium and iodide of potassium.

Principal Preparations:—

Tincture of Iodine, 8 per cent. in alcohol. Dose, ℥j–v.

Compound Tincture of Iodine (iodine, 5 per cent., potassium iodide, KI, 10 per cent., and water 85 per cent). Dose, ℥j–x diluted.

Potassium Iodide. Dose, gr. v–xxx.

Iodoform, CHI_3. Dose, gr. j–v, in pill form.

Actions.—Iodine in its elementary state is an irritant to the skin, and is much used in the form of tincture to produce counter-irritation.

Internally in small **doses** it is stimulant and tonic; it excites a sensation of heat or burning in the stomach, and in large doses acts as an irritant poison. If continued for any length of time, iodine induces great waste and rapid elimination of waste products, causing anæmia and depression.

The local irritant effect is diminished when combined with potassium; hence, potassium iodide (aqueous solution of potassa and iodine) is usually employed for internal use, which allows the administration of larger doses and for a greater length of time.

Toxicology.—*The antidote* for iodine is starch, forming an iodized starch, which should then be evacuated from the stomach.

Colorless Iodine.—There are a number of methods for bleaching iodine; among them are the following: 1st. Add to a drachm of tincture of iodine six ounces of hot water and twelve grains of phenol; stir with a glass rod. 2d. Iodine is bleached by mixing with carbolic acid; this, carbolate of iodine, combines all the advantages of both agents.

Dental Uses.—Iodine is a very valuable agent in dental practice, the tincture being employed locally in the treatment of periostitis, inflammation and ulceration of the gums, fungous growths, suppurating pulps of teeth, alveolar abscess, and for ulcerations of the mucous membrane; it is often combined with carbolic acid, and for dental periostitis it is generally combined with tincture of aconite; this combination forms an excellent treatment for the incipient stages of this affection, as well as those of alveolar abscesses.

TURPENTINE.

Derivation.—Turpentine is a concrete, oleo-resinous exudation from various species of pine, but principally from the "yellow pine."

Properties.—Turpentine is in the form of tough, yellowish masses, more or less transparent, inflammable, having a strong, unpleasant odor, and warm, pungent taste.

It is composed entirely of resin and the essential oil known as oil of turpentine, $C_{10}H_{16}O$. It is soluble in alcohol. The oil is the form mainly used.

Actions.—Turpentine is a stimulant, diuretic, antispasmodic, and rubefacient (counter-irritant), and antiseptic externally.

Principal Preparations:—

Oil of Turpentine, Spirits of Turpentine, a volatile oil distilled from turpentine. Dose, ♏ v-xv in emulsion as a stimulant.

Turpentine Liniment.—Resin cerate (a composition of wax, oil or lard) 65 per cent., oil of turpentine 35 per cent.

Pitch is a resinous exudation from the stem of certain pine, fir, and spruce trees. It melts at the boiling point of water, and softens by the heat of the human body. It is of a dark brown color, and possesses a well-known odor and taste. It is used principally as the base of plasters.

Dental Uses.—Turpentine may be used in dental practice for its rubefacient and antiseptic properties.

ASTRINGENTS.

Astringents are agents which produce contraction and condensation of organic tissues, with a tendency to remove morbid affections, arrest hemorrhage and excessive secretions from the mucous membrane. They are divided into two classes, known as vegetable and mineral.

The **principal vegetable astringents** are tannic acid and gallic acid, the chief element of these being tannin, *while the mineral astringents* are persalts of iron (see Tonics), alum, sulphuric acid, nitric acid, etc.

TANNIC ACID, $C_{27}H_{22}O_{17}$.

Derivation.—Tannic acid is obtained from nut galls (galls of the Dyer's oak). *Galls* are the excrescence on the twigs of the Dyer's oak, grown in Asia Minor and Persia, caused by the punctures and deposited ova (egg) of an insect.

Properties.—Tannic acid is obtained in the form of thin, yellowish crystals, inodorous, very soluble in water, less so in alcohol and ether.

Actions.—Tannic acid is the most powerful of all vegetable astringents and styptics. It is especially active upon albumin, gelatin, and fibrin, forming therewith insoluble tannates, thus protecting the parts beneath until resolution occurs. Dose, gr. j-xx, in pill.

Dental Uses.—Tannic acid is a very valuable agent to the dental practitioner. It is used locally in the treatment of hemorrhage following the extraction of teeth, wounds of the mucous membrane, fungous growth of the tooth pulp, hypertrophy of the gums, and to many it has proven beneficial in the treatment of sensitive dentine, a strong solution of tannin being mixed with alcohol. *In mercurial salivation*, the powdered tannic acid, moistened with water, will check the tendency to absorption and the consequent loosening of the teeth, and will render the gums firmer and more comfortable.

Glycerite of Tannic Acid (tannin, ℥ij ; glycerin, ℥viij), for external use.

Ointment of Tannic Acid (tannin, ʒj ; lard, ℥j), for application to ulcers, etc.

GALLIC ACID, $C_7H_6O_5$.

Derivation.—Gallic acid is prepared from nutgalls. The powdered galls, in water, are left to the action of the atmosphere, when the acid, in the form of fine, almost colorless, crystals are deposited.

Properties.—Gallic acid is obtained in the form of very fine, silky, and almost colorless crystals. It is slightly soluble in cold water (100 parts), and rapidly so in hot water, glycerine, or alcohol. It has a slightly acid and astringent taste.

Action.—Gallic acid is a powerful astringent, styptic, and disinfectant. It is given directly for internal hemorrhage, profuse perspiration (night sweats), and excessive expectorations of phthisis, and chronic diarrhœa. Dose, gr. v-xx, in pill form.

Dental Uses.—In dental practice gallic acid may be used as a styptic in superficial hemorrhages; it is not as efficient, however, as tannic acid. It is employed in the form of a gargle, in acute inflammation of the mucous membrane, etc.

ALUM.

Source.—Alum is found native in Italy, in the neighborhood of volcanoes. It is also obtained from aluminous slate or shale by roasting and exposure to the air.

Formula.—The official alum (potassic-aluminic-sulphate) has the formula $K_2AL_24SO_4 + 24H_2O$. Dried or "burnt alum" has the water of crystallization, $24H_2O$, driven off by gentle heating, which leaves it in the form of a soft, white powder.

Properties.—Alum is a white, transparent salt, crystallizing easily in octahedrons (having eight equal and equilateral triangles). It dissolves easily in hot water, and by about fifteen times its weight in cold water; is insoluble in alcohol. It possesses an astringent and sweetish taste.

Actions.—When taken internally in large doses, it causes vomiting, purging, and inflammation of the gastro-mucous membrane. As an emetic, powdered alum, in teaspoonful doses, is very efficient. Applied locally, it is an excellent astringent to relaxed or bleeding parts. Dose, in powder or solution, gr. x–xl (2ʒ).

Dental Uses.—In dental practice, alum is employed as a styptic in alveolar hemorrhage, superficial hemorrhage of the mucous membrane, ulcers of the mouth, etc. It also serves an excellent purpose as a gargle in ulceration and sponginess of the gums.

STYPTICS AND HÆMOSTATICS.

Styptics are agents which arrest hemorrhage by local application. They are divided into chemical and mechanical, according to their action.

Chemical styptics coagulate the exuding blood, and at the same time stimulate the tissues to contraction.

The principal **members** of this class are, tannic and gallic acids (see Astringents), persulphate of iron—solution, subsulphate of iron—powdered (see Tonics), and alum (see Astringents).

Mechanical styptics are agents which promote clot formations in the mouths of bleeding vessels. They retard the flow by detaining the blood in their meshes, or absorb it until it coagulates.

The principal mechanical styptics are spider's web, plaster-of-paris, sandarach varnish, cotton, etc.

Hæmostatics are agents capable of arresting hemorrhage by internal administration, such as ergot, antipyrine (see Astringents), the diluted mineral acids (see Tonics), etc.

ERGOT.

Source.—Ergot is obtained from a parasitic (growing upon or within another, a hanger on) fungi replacing the grain of rye. It is a diseased state of the grain, occasioned probably by a hot summer succeeding a rainy spring. *Corn ergot* is obtained from a similar growth upon the Indian corn.

Principal Preparations:—

Fluid Extract of Ergot. Dose, ʒss–ij.

Extract of Ergot. Dose, gr. j–xx.

Medical Properties and Actions.—Ergot is a hæmostatic, aiding coagulation by slowing the blood current. It is also used to stimulate the contraction of unstriped muscular fibre, particularly those of the uterus, causing continuous labor pains. It has been much used for this purpose in obstetrics, and very often injuriously, causing laceration of the perineum and paralysis of the fœtal heart, the natural intermitting contraction being the most desirable.

Dental Uses.—Ergot may be used in dental practice for its hæmostatic properties in alveolar or other hemorrhages.

ESCHAROTICS OR CAUSTICS.

Escharotics or caustics are agents which are capable of destroying the life of the tissue with which they come in contact, producing an eschar or sloughing of the tissue. Fire itself is the actual cautery, while the potential cautery (caustic substances) is represented by silver nitrate, arsenious acid (see Tonics), carbolic acid, zinc chloride (see Tonics), and the mineral acids (see Tonics), etc.

NITRATE OF SILVER, AgNO₃.

Derivation.—Nitrate of silver, or "lunar caustic," is made by dissolving silver in nitric acid, and evaporating the solution. The reaction being $Ag_3 + 4HNO_3 = 3AgNO_3 + 2H_2O + NO$.

Properties.—Nitrate of silver is in the form of colorless, shining crystals, but is readily blackened by mixing with organic matter or by exposure to the light, is very soluble in water, and has a strong metallic and styptic taste. It is often cast in sticks, by first being melted (fusing at 426° F.) and then poured into suitable moulds.

Actions.—Nitrate of silver is a powerful caustic and astringent, a heart and nerve stimulant, antispasmodic and sedative. When applied locally to the mucous membrane, ulcers, etc., it first turns the surface white, owing to its union with the coagulated albumin, but finally turns to a black color, which is due to the partial reduction of the silver by the sulphuretted hydrogen contained in the atmosphere. *Continued use* of nitrate of silver will cause a peculiar blue line in the gums, similar to that from lead poisoning; this is followed by a blue appearance of the skin. The remedy should be discontinued at once when this discoloration is observed.

Dose of nitrate of silver, gr. ⅙ gradually increased to gr. j, in pill form. Never should be given with tannin or a vegetable extract; an explosive compound may result. The fused or solid form is used externally.

The antidote for nitrate of silver is chloride of sodium (common salt) freely; it precipitates it in the insoluble chloride of silver; also acts as an emetic.

Dental Uses.—Nitrate of silver is employed in dental practice for obtunding sensitive dentine, especially where the cause is mechanical abrasion, or from the fracture of a tooth, exposing the healthy and sensitive dentine, the stick form being employed, or the end of a silver wire may be immersed in nitric acid and carefully applied. It is also a valuable application for ulcerated conditions of the mucous membrane of the mouth, also as a treatment for caries in deciduous teeth.

CARBOLIC ACID, C_6H_5O.

Derivation.—Carbolic acid, phenylic alcohol or phenol, is obtained as an alcoholic product of the distillation of coal tar, between the temperatures of 338° and 370° F. Carbolic acid, though the universal name, is inappropriate. It does not belong to the acid series (it will not turn blue litmus paper red), being neutral in its reaction.

Properties.—Carbolic acid, when pure, is in the form of colorless or pinkish acicular (needle-like) crystals. It becomes an oily liquid at 95° F., or, if exposed to the air, the crystals readily absorb moisture and are thus liquefied. Five per cent. of water liquefies it; any further addition simply forms a mechanical mixture. It is freely soluble in alcohol, ether, chloroform, glycerine, and the essential oils. It has a strong aromatic odor and taste, resembling creasote somewhat.

Actions.—Carbolic acid in its pure state is escharotic; when diluted, it is a powerful antiseptic, germicide, rubefacient, and is a violent poison; internally, it is a sedative and carminative, allaying vomiting and gastric irritability.

It resembles creasote closely in many of its medicinal properties, but is probably more efficacious, and its odor is surely less objectionable; the use of creasote by the dental practitioner may therefore well be discarded.

Dose, gr. ¼, for relief of nausea, etc.

Dental Uses.—Carbolic acid is a valuable agent in dental therapeutics, it being one of the best escharotics, styptics, antiseptics, sedatives, etc. It is used to obtund sensitive dentine, to relieve odontalgia, when caused by the exposure of the tooth pulp, by applying it to the exposed surface; it arrests putrefactive changes, is a valuable agent in the treatment of alveolar abscess; is also used to bathe cavities in the teeth, both for its obtunding effect upon the sensitive dentine and to destroy any low organisms that may be in the softened dentine.

In a **form known as phenol sodique**, carbolic acid is very useful as a styptic for the treatment of superficial hemorrhage after the extraction of teeth, and forms an excellent antiseptic mouth wash.

"Combined with glycerine (1 part to 12 of glycerine) it will stimulate the mucous secretion, and hence has been applied to the palate, in cases of deficiency of this secretion, to promote the suction of upper dentures."*

ANTIZYMOTICS.

Antizymotics are agents which arrest or prevent fermentative processes; they are divided into antiseptics and disinfectants.

Antiseptics are those agents which prevent or retard septic decomposition, either by destroying the bacteria upon which putrefaction depends or by arresting their development.

The most important of this group are bichloride of mercury, peroxide of hydrogen, carbolic acid (see Escharotics), potassium permanganate, iodoform, phenol sodique, alcohol, eucalyptol, etc.

Disinfectants are those agents which destroy the germs of infectious diseases.

The principal members of this group are carbolic acid (see Escharotics), zinc chloride (see Tonics, Zinc), potassium

---

* Prof. Gorgas.

permanganate, iodine (see Irritants), aromatic sulphuric acid (see Tonics,—Sulphuric Acid).

BICHLORIDE OF MERCURY, $HgCl_2$.

Derivation.—Bichloride of mercury, mercuric chloride, or "corrosive sublimate," is obtained by distilling* a mixture of sodium chloride and mercuric sulphate; a double decomposition takes place, forming mercuric chloride and sodium sulphate.

Properties.—Bichloride of mercury is in the form of colorless crystalline masses. It is inodorous, fusible, soluble in 16 parts of water, 7 parts of alcohol and ether, and has an acrid, styptic taste.

Actions.—Bichloride of mercury is one of the most active salts of mercury. It is one of the most efficient of all the antizymotics in the strength of 1 part to 2000 parts of water. It is internally employed in chronic diarrhœa, dysentery, and syphilis.

Dose, gr. $\frac{1}{30}-\frac{1}{16}$ in pill form.

Antidotes to bichloride of mercury are albumin, wheat flour, milk, etc.

Dental Uses.—For prophylactic treatment of the oral cavity, particularly of the teeth, bichloride of mercury, 1 to 10,000, is most effective. It should be carefully used, however, on account of its poisonous character.

It is also used extensively in dental practice, in treatment of alveolar abscesses, and in diseases of the antrum of Highmore, in a solution of 1 to 2000, to 1 to 5000.

PEROXIDE OF HYDROGEN, H_2O_2.

Derivation.—Peroxide of hydrogen is obtained by combining an extra molecule of oxygen with hydrogen monoxide,

---

* The double process of vaporization and condensation of the vapor.

H_2O (water), the result being a water-like liquid, H_2O_2. "As when barium dioxide is dissolved in dilute hydrochloric acid:—

$$BaO_2 + 2HCl + H_2O = BaCl_2 + H_2O + H_2O_2."*$$

Properties.—Peroxide of hydrogen is in the form of a colorless, transparent liquid, is inodorous and almost tasteless.

Actions.—Peroxide of hydrogen is one of the most efficient and at the same time the least harmful of all antiseptics and disinfectants. The second molecule of oxygen, spoken of above, is very loosely combined, and the mixture is always on a strain to break up into water and oxygen; for this reason it should always be kept in a cool and dark place, and it is owing to this fact (that peroxide of hydrogen generates "ozone," O_3) that pus and the bacteria of diseased surfaces, when treated with this agent, are at once destroyed. "As soon as ozone has accomplished its cleansing effects upon the infected surface, it is readily transformed into ordinary oxygen, owing to its instability." *It is employed as an internal remedy* in fevers, whooping cough, bronchitis, consumption or phthisis, diphtheria, dyspepsia, catarrh of the stomach, etc.

Locally Employed.—Peroxide of hydrogen may be employed for its antiseptic and pus-destroying properties in the treatment of abscesses, ulcers, carbuncles, wounds, both fresh and putrid, catarrh of the nose, hay fever, diphtheria, etc. It is also the base of most hair-bleaching solutions.

Dose of peroxide of hydrogen, ʒss–ij.

Dental Uses.—It is one of the most valuable remedies in dental therapeutics; it is especially valuable in the treatment of alveolar abscesses, pyorrhœa alveolaris, inflammation and ulceration of the oral mucous membrane, fungous growths, etc.

---

* Leffmann's "Chemistry."

We give below the result of comparative tests made by Chas. Marchand, chemist, of New York, demonstrating experimentally, as he says, the difference between the bactericide potency of the following chemicals. The experiments were made upon a diphtheritic membrane. We quote his list, starting with peroxide of hydrogen :—

QUANTITY OF THE MIXTURE OR SOLUTION REQUIRED TO ANNIHILATE MICROBES.

	Cubic Centimetres.
Marchand's Peroxide of Hydrogen, medicinal (*harmless*),	2.00
Bichloride of mercury,	3.00
Nitrate of silver,	5.00
Hypochlorite of lime,	9.00
Chlorine gas (aqueous solution),	10.00
Iodine,	10.00
Bromine,	24.00
Iodoform (when fresh),	28.00
Salicylic acid,	40.00
Muriatic acid,	100.00
Carbolic acid,	128.00
Permanganate of potash,	140.00
Chlorate of potash,	158.00
Alum,	180.00
Tannin,	190.00
Common salt,	196.00
Sulphide of calcium,	201.00
Boracic acid,	300.00
Sulphurous acid,	325.00
Lactic acid,	360.00
Chloride of iron,	371.00

By referring to the tests published above, it is readily seen that peroxide of hydrogen, as a bactericide, is 1½ times as strong as bichloride of mercury, 2½ times as strong as nitrate of silver, 5 times as strong as iodine, 14 times as strong as iodoform, 20 times as strong as salicylic acid, 64 times as strong as carbolic acid.*

PERMANGANATE OF POTASSIUM, $K_2Mn_2O_8$.

Derivation.—Permanganate of potassium is prepared by fusing the black oxide of manganese with chlorate of potassium and caustic potassa.

Properties.—It is in the form of dark purple crystals; it

---

* *The Dangers of Carbolic Acid.*—The following experiments, made by Mr. Marchand, show the dangers of applying carbolic acid in the treatment of suppurative diseases.

Six dogs were submitted to the action of this corrosive antiseptic in the following manner:—

Two to three square inches of hair on the leg of each animal upon which the experimentation took place was closely shaved.

Morning and evening an application, from 8 to 10 drops (three per cent. solution) of carbolic acid was made upon these prepared surfaces, and continued for ten days.

One hour or so after each application the surface was dry, owing to the evaporation of water, and then, as an immediate consequence, it was covered by a small quantity of pure, concentrated carbolic acid, of which the corrosive properties are well known.

At the expiration of said time two of these dogs were sick, each one having an ulcer on the prepared surface of the leg, which was due to the repeated application and evaporation of the three per cent. solution of carbolic acid; and three days later each one of the four other dogs had an ulcer of the same nature, which was produced from the same cause.

These four were then submitted to treatment by Peroxide of Hydrogen, which in four weeks thoroughly cured them.

The application of the three per cent. solution of carbolic acid on the ulcers of the two other dogs was again continued, and on the fiftieth and sixty-second day, respectively, both animals expired from blood-poisoning. The autopsy showed that the blood of these animals was invaded by the bacteria of Davaine, which were detected by a microscopical examination.

is very soluble in water, forming a beautiful lilac-colored solution; inodorless, and has a sweetish, astringent taste.

Actions.—Permanganate of potassium taken internally is a stimulant, and is given with benefit in dyspepsia; is a mild escharotic, and a powerful disinfectant and deodorizer. The solution is decomposed by organic matters, sulphides and sulphites, yielding up its oxygen readily (on which property its use depends), and is converted into a colorless solution.

The most important uses for this agent are externally, where it is employed as a deodorizer and disinfectant in abscess, ulcers, cancers, caries of the bone, etc., in the form of a lotion and spray, while the powder may be sprinkled on gangrenous surfaces, acting as a local stimulant as well as a deodorizer.

Dose, for internal use, gr. ¼–j. For external use, fʒj to water, fʒv–x.

Dental Uses.—It is employed in dental practice in the treatment of foul abscess, in diseases of the antrum, necrosis of the maxillary bones, ulcers of the mouth attended with fetid discharges, offensive breath, etc.

IODOFORM, CHI_3.

Derivation.—Iodoform is a preparation of iodine. It is "obtained by the action of chlorinated lime upon an alcoholic solution of iodide of potassium, heated at 104° F., the product being iodoform and iodate of lime, the iodoform being separated by boiling alcohol."

Properties.—Iodoform is in the form of small yellow crystals, which are soft to the touch, of a sweetish taste; is volatile, and has a very unpleasant odor (which may be covered with oil of rose, etc.); is insoluble in water, but is soluble in alcohol, ether, chloroform, and the essential oils.

Actions.—Iodoform is an efficient antiseptic, inhibiting and destroying the microbes of putrefaction and suppuration ; it has also slight local anæsthetic properties.

Internally in small doses it is a tonic, alterative, anodyne, and antiseptic.

In large doses it causes a form of intoxication, followed by convulsions, collapse, and death.

Dose, gr. j-v, in pill form.

Dental Uses.—In dental practice iodoform is a valuable agent; it is an excellent antiseptic for the treatment of alveolar abscesses, putrescent pulps, especially when combined with oil of eucalyptus. Prof. Peirce recommends iodoform ground with equal parts of oil of cloves and oil of eucalyptus, a portion of which may be introduced to the inflamed part on the point of a small broach.

Iodoform is also serviceable as a packing for the pockets of pyorrhœa alveolaris, and as a dressing or packing for wounds, where it may be used in the powdered form or in the form of a gauze which is prepared for the purpose.

ARISTOL.

Derivations.—Aristol is a preparation of iodine, and has gained a position in recent therapeutics as a perfect substitute for iodoform. It is a thymol iodide, and is " produced by treating an aqueous solution of iodine in iodide of potassium with an aqueous solution of thymol in the presence of caustic potash."

Properties.—Aristol is in the form of a reddish brown powder; is volatile, and has a slight aromatic odor, though when compared with iodoform we may say it is practically inodorous. The quantity of iodine contained in it is from 45 to 50 per cent.

Aristol is insoluble in water and glycerine, slightly soluble in alcohol, but is readily soluble in ether, chloroform, collodion, and the fixed and ethereal oils.

Actions.—Aristol is one of the most efficient antiseptics; it is particularly applicable as a dressing for wounds, ulcerations, and abrasions of the skin and mucous membrane.

It may be dusted over the wound, or applied in the form of aristol ointments (compounds of vaseline or cold cream) or in combination with collodion. All of these are excellent and easy of application. Its efficiency as an antiseptic and alterative is largely due to the fact that it slowly gives off its iodine, and it is also due to this fact that it does so decompose, when exposed to the light or undue heat, that it should be kept in a colored bottle or a closed box and in not too warm a place. If this is not done the loss of iodine will be readily noticed by it gradually becoming paler in color.

Dental Uses.—Aristol is a welcome addition to our catalogue of therapeutic agents. I have found it an excellent agent for treating root canals from which putrescent pulps have been removed and for alveolar abscesses. It can also be used advantageously in combination with root-filling materials; it may be used by mixing the powder with chloro-percha, or where gutta-percha cones are to be used the cone may be dipped in a solution of aristol and chloroform and immediately carried to position. Dr. Kirk says of it in this connection: "I have made use of aristol in connection with root-filling materials by another method. A strong solution of aristol is made in the oil known to house-painters as 'Japan dryer,' sufficient of the drug being added to make the liquid somewhat thinner than glycerine; into this is worked with a spatula freshly calcined oxide of zinc until the mass is like putty, in which condition it is to be worked in to the root-canals. The mass becomes quite hard, and seems to fulfill admirably the requirements of a root-filling."

CATHARTICS.

Cathartics or purgatives are agents which hasten the intestinal evacuations; they comprise such substances as magnesia preparations (Epsom salt), senna, rhubarb, fruits of various kinds, etc.

MAGNESIA, MgO.

Derivation.—Magnesia or magnesium oxide is obtained by subjecting magnesium to a red heat in the open air, when it will burn with a bright light and produce MgO.

Properties.—Magnesia is a very light, white powder, odorless, has an earthy taste, is freely soluble in water, more so in cold water, and neutralizing acids.

Actions.—Magnesia is an efficient aperient (mild purgative), is antacid, hence an excellent remedy for great acidity of the stomach, and is the antidote for poisoning by mineral acids. When it is desirable to administer magnesia in large doses, and for a considerable length of time, it may be given in connection with lemonade, which will render it more soluble, avoiding its accumulation in the bowels.

Chief Preparations:—

Sulphate of Magnesia, "Epsom salts," "salts." Dose, ʒj-ʒj, in water, a popular purgative.

Liq. Citrate of Magnesia (magnesium carbonate), citric acid, potassium, bicarbonate and water. Dose, ℥iv-vj.

Dose of Magnesia as an aperient, gr. x-ʒj. As an antacid, gr. 20 (Ə j).

Dental Uses.—It is employed in dental practice as an ingredient of dentifrices, for its antacid properties, and in solution as a mouth wash or gargle, to counteract the effect of certain medicines (acids) upon the teeth.

APPENDIX.

EMERGENCIES.

PRELIMINARY REMARKS.

Nothing, of course, can take the place of the advice and service of an experienced physician in time of emergencies; but the physician is not always at hand, and accidents of various kinds may occur in the dental office, or patients in distress present themselves to the dental practitioner for immediate relief, or some one may be burned, cut, poisoned, suffocated, or drowned, where, if we possess presence of mind, and sufficient knowledge, it may be our privilege to save an endangered life. It is, therefore, expedient that the dental student should have a more thorough knowledge of what to do in case of such emergencies; it is for that purpose that this chapter of practical suggestions is prepared.

Apoplexy is the rupturing of a blood-vessel in the brain. The symptoms are stupor, heavy snoring breathing, slow pulse, flushed face, followed by paralysis usually of one side, this being marked by the drawing up of one side of the face.

Treatment.—Loosen clothing about the neck, make cold applications to the head, and keep the patient in a sitting posture until the physician arrives.

Burns or Scalds.—Not unfrequently does some one's clothing take fire, usually that of women, on account of the character of their clothing. The first thing to do in time of such an accident is to have the patient lie down, but if she loses her presence of mind and will not obey instructions, throw her down, then cover or envelop her at once with the first article you seize that will exclude the air and smother the flame, a breadth of carpet, rug, blanket, coat, etc.

After the fire is extinguished, or after an extensive scald, if there is much of a burn or blister, the clothing, as much as needs be removed, should be carefully clipped away, so as not to break the blisters that may have formed. These may be punctured at one edge and their contents discharged, when the outer skin will fall back in place. Then a dressing of pure sweet oil or castor oil should be carefully applied on strips of soft linen. When the skin is destroyed, the air may be excluded by applying at once, either of the following: sweet oil, linseed oil, collodion, vaseline, etc. Dr. Charles Dulles, in his manual on accidents and emergencies, says: "In case of a person severely and extensively burned, the entire body may be immersed in a bath, which shall be kept, as long as necessary, at a temperature of 100°. Where the shock of a burn is great, some stimulant should be given, and laudanum in twenty drop doses to an adult, and half as much to a child, to allay the suffering."

For Slight Burns or Scalds, an excellent dressing is to quickly sprinkle the parts with bicarbonate of soda and cover same with wet cloth, or they may be painted with white lead or covered with the white of an egg or carron oil (equal parts of linseed oil and lime water)—in fact, anything that will exclude the air and prevent friction, and will not prevent after examination, may be used. Aristol ointment (aristol in cold cream or vaseline) is also being used with good results, while for small burns on the hand, arm, leg, etc., immerse instantly in cold water, and let it remain for some length of time.

Burns with Acids or Caustic Alkalies, such as soap lye, should be deluged with water, and followed by an application of bicarbonate of soda for the former and vinegar for the latter, to be followed by an application of oil.

Catalepsy in appearance somewhat resembles death. The patient becomes unconscious, the muscles rigid, and the skin pallid. In itself it is by no means dangerous, and it affords time enough to summon a doctor, which is the only sensible thing to do under these circumstances.

Choking is usually caused by the lodgment of some foreign substance in the trachea or œsophagus. When the body is lodged in the trachea there is great irritation and coughing, though it does not materially interfere with deglutition. While, on the other hand, when the œsophagus is closed, it is usually impossible to swallow, and there is little or no coughing.

Treatment.—Hold the head low and slap the back quite forcibly. Blow into the ear, which will excite a reflex action that will aid the patient in expelling the foreign body. The removal of pins, needles, splinters, fish bones, etc., from the throat is usually an extremely delicate operation. They should be grasped with a small pair of forceps or tweezers, or a blunt pair of scissors may be used for the same purpose.

Convulsions are usually caused by some irritation of the digestive apparatus, or *by some interference in the eruption of the teeth.*

Treatment.—When the physician's coming is delayed, the child should be placed in a hot bath; the head at the same time should be kept cool by cold applications. This should be continued for about ten minutes, when the child should be wrapped in warm blankets and put to bed. *If there should be one or more teeth endeavoring to erupt at this time, the gums should be freely lanced.*

Dislocations can be easily detected. There is always deformity, pain, and stiffness of the joint affected.

Dislocation of the lower jaw with treatment for same, is fully treated upon pages 55 and 56.

Dislocation of the fingers can usually be corrected by strong pulling and at the same time pressing the parts into place, where they should be retained for several days by a splint and bandage.

Dislocations of other joints had better be left for the surgeon's hands. "The risk of doing injury by injudicious efforts to set a joint is greater than that of waiting until a surgeon can be summoned." The patient, however, should be placed in the most comfortable position and hot fomentations should be applied.

Drowning.—It is important to remember that the body, as a whole, is a very little lighter than water, therefore, a person who is in danger of drowning should lie flat on the back and keep the entire body, with the exception of the mouth and nose, under water. The arms should be stretched at full length above the head, and the lungs should be kept filled with air as much of the time as possible. This would very materially aid both the one in danger and the rescuing party.

Resuscitation.—*Avoid delay.* Do not wait to carry the patient to a house or hospital, but treat him on the spot. "Remember that the patient is suffering from two things, want of air or oxygen and loss of heat from the body." The first thing to do, then, is to free the body from any clothing that may interfere with respiration,—that is, about the neck, chest, and waist. If natural breathing has ceased, artificial respiration should be commenced as soon as possible. First, hastily make a roll from clothing, blankets, or anything that may be at hand, place the patient over this, *face downward*,

134   APPENDIX.

Fig. 18.

Forcing Water and Mucus out of the Lungs and Throat.

allowing his forehead to rest upon one hand to keep the mouth and nose clear of the ground. Place the hands, well spread, upon the patient's back, over the stomach and base of thorax. Then with a forward motion throw all the weight upon them that the age and sex of patient will justify. Repeat this three or four times, which will cause the water and mucus to run out of the mouth, throat, and trachea. (See Fig. 18.) Wrap a handkerchief around the forefinger and pass it into the mouth and remove any mucus that may remain. Turn the patient on his back, grasp the tongue and draw it forward and down on to the chin, lay a strip of the handkerchief or other material across the tongue and pass the ends behind the neck and tie, or have some one to hold the tongue to keep it from falling back and closing the throat. Then begin artificial respiration.

Howard's Method.—The patient is placed upon his back, his arms extended backward and outward, where they should be held by an assistant. A roll of something (clothing, a folded blanket, coat, or stick of wood) is then placed under the false ribs so as to throw them prominently forward. The operator should then kneel astride the patient's abdomen, placing both hands so that the fingers will press into the intercostal spaces on each side, and the base of the thumb rest upon the anterior margin of the false ribs. The operator should then place his elbows firmly against his sides, and throw himself forward, bringing his weight to bear upon the patient's false ribs, forcing them inward and upward toward the diaphragm, then suddenly let go and return to the erect position. Repeat these movements ten to twelve times per minute until natural breathing begins, which will gradually take the place of the artificial. Fig. 19 illustrates this method.

Sylvester's Method.—After the patient has been placed

136   APPENDIX.

Fig. 19.   Artificial Respiration. Howard's Method.

upon his back, with folded clothing under his shoulders, the operator should kneel behind his head and go through the following manipulations:—

First, to induce inspiration: Grasp both arms just below the elbows and swing them around horizontally until they nearly meet above the head, with the back of the hands or elbows touching the ground; hold them there for three or four seconds. This draws the ribs up so as to expand the chest and allows the air to enter the lungs. (See Fig. 20.)

The second movement is to induce expiration. Bend the arms at the elbows, and carry them down so that they rest upon either side of the chest. Bring the weight of your body upon them, pressing forcibly and steadily, which pressure, if continued for a few seconds, will force the air out of the lungs. (See Fig. 21.) These movements should be continued alternately twelve to fifteen times per minute.

When natural breathing is attempted it may be stimulated by applying smelling salts or ammonia to the nose, by slapping, or by pouring hot and cold water alternately upon the chest. When the patient is able to swallow, some stimulant should be given every few minutes until the danger point is passed, such as a teaspoonful of whisky or brandy, or double the quantity of hot water. After the patient is resuscitated he should be wrapped in warm blankets and carefully carried, with the head low, to a warm bed.

Epileptic Fits are characterized by sudden loss of consciousness and power of coördination of motion; there is a rigidity of motion which is followed by violent convulsions of a short duration, usually accompanied by more or less foaming at the mouth. There is also a peculiar cry that accompanies these attacks, caused by laryngeal spasms.

Treatment.—There should be no struggling with the patient, but an effort should be made to regulate the movements so

138   APPENDIX.

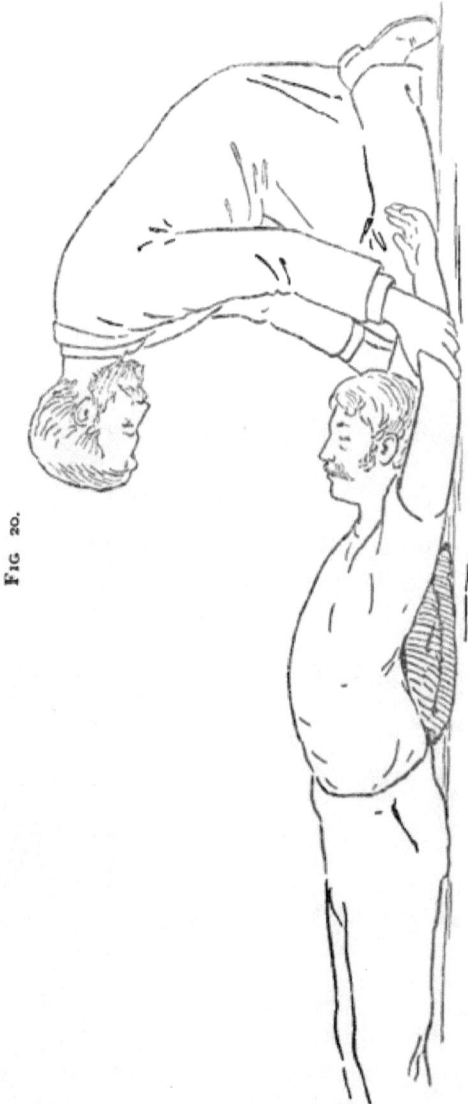

Fig. 20.

Artificial Respiration—Sylvester's Method. First Movement—Inspiration.

APPENDIX.

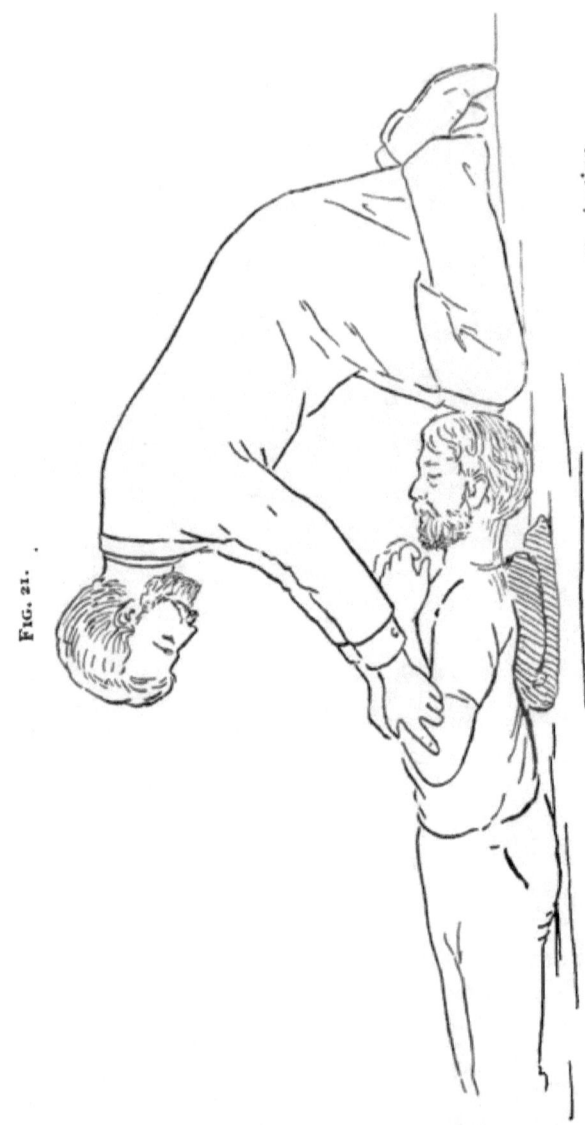

Fig. 21.

Artificial Respiration—Sylvester's Method. Second Movement—Expiration.

that the patient will not do himself any harm. A folded napkin or towel or a soft piece of wood should be placed between the teeth to prevent biting the tongue. As soon as the convulsions are passed, the patient should be allowed to rest quietly in bed for some time. Dr. Dulles says:—"It would be a good plan if every one who is subject to epileptic attacks had his or her name and address placed inside the coat or in some place where it could be seen at once when the clothing is loosened to give relief, as is almost invariably done when such attacks occur. Epileptics should not, except where it is absolutely unavoidable, go about alone, or go into crowded places. They have no right, on their own account and for the sake of others, to incur risks involved in such conduct, except under the stress of necessity."

Exhaustion, Heat.—This is not a serious illness. It is usually caused by physical overwork in hot and badly ventilated rooms or in the heat of summer,—the latter is apt to be confounded with heat or sunstroke. In heat exhaustion, however, there is a cool, moist skin, while in sunstroke the bodily temperature is raised and the skin is quite dry.

Treatment of heat stroke is very simple. The patient should have complete rest, plenty of fresh air, and a mild stimulant—hot soup, milk, tea, or coffee, or a small dose (one or two teaspoonfuls) of wine or brandy.

Fainting is caused by the cutting off of the supply of blood to the brain, which is due to a temporary weakening of the heart's action.

Treatment.—If, for any reason, the patient does not fall to the floor or couch so as to bring the head as low as the body he should be placed in such a position. This is to aid the heart in sending the blood to the brain, and if the head is placed a little lower than the body, the force of gravity will aid very much in sending the blood where it is needed. At the same time the clothing should be loosened and plenty of fresh air admitted. Cold water may be sprinkled upon the face, and ammonia or smelling salts applied to the nose tend to excite the nerves of sensation and rouse the heart to renewed activity.

Fractures.—Fractured bones may be recognized by the deformity and abnormal mobility of the parts affected. It is also usually accompanied by crepitus,—a clicking or grating sensation at the point of fracture.

The Treatment of fractures should not be attempted by any one who has not surgical training, and as there is no great haste necessary, an effort should be made to simply place the patient in as comfortable a position as possible until the physician arrives. The following notes and illustration (Fig. 22) are taken, with permission of the author, from Dr. G. R. Butler's " Emergency Notes:"—

"*If patient must be moved, or travel some distance*, apply temporary splints or dressings to prevent further injury from movement.

"In city or country one of the following list of materials may usually be found, and, by the aid of a little ingenuity, utilized for temporary dressings:—

"*For Splints.*—Cigar boxes, pasteboard (boxes, book-covers), folded newspapers, shingles, barrel-staves, umbrellas, walking-sticks, rulers, paper-cutters, branches, twigs, straw, fence-palings, spoons, tongs.

"*For Padding.*—Cotton, flannel, towels, flax, jute, oakum, hay, moss, piece of blanket, coat-sleeves or stockings stuffed with grass, hay, straw, or leaves.

"*For Bandages.*—Handkerchiefs, stockings, garters, suspenders, sheets, blankets, and patient's clothing torn into strips.

"The temporary treatment of the following special fractures

142   APPENDIX.

Fig. 22.

Improvised Splints and Dressings for Fractures of Jaw, Upper Arm, Forearm, Thigh-bone, and Leg.

should be studied and personally practiced. Some of these improvised splints and their application are shown in Fig. 22.

"*Jaw-bone*.— Four-tailed bandage. (See page 42.)

"*Collar-bone*.—Broad bandage around chest and arm of injured side. Broad arm-sling.

"*Upper Arm-bone*.—Two splints. Narrow arm-sling.

"*Forearm and Wrist bone*.—One splint reaching to base of fingers. Broad arm-sling.

"*Ribs*.—Broad bandage around chest, applied during expiration—*i. e.*, chest emptied of air.

"*Thigh-bone*.—Preferably a long, posterior (back) splint, reaching from waist to heel. If this cannot be secured, simply tie limbs together.

"*Leg and Ankle-bones*.—Internal (inside) splint, or tie limbs together." See Fig. 22 for illustrations.

Freezing of the ears, nose, hands, or feet sometimes takes place in such a deceptive manner that much damage is done before it is recognized, for one may go sometimes with a part of the body frozen and not be aware of it, on account of the absence of sensation in the part, which is due to the freezing of the nerves.

Treatment.—Warmth and circulation is the first demand, and this should be restored gradually until normal (98° Fahrenheit). This is best accomplished by applying snow or cold water to the part, to which gradually add warmer water; gentle friction, too, is found beneficial. The after effects, if serious, should have similar treatment to that of a burn. If the whole body or a large part of it is frozen, after the normal temperature is restored the patient should be wrapped in warm blankets and given some internal stimulant, as tea or coffee, until the physician arrives.

Hemorrhage.—There is no accident where a little accurate knowledge and a level head is of more value than in case of hemorrhage, whether it be from some external wound or from the rupture of a blood-vessel internally. Hemorrhage may occur from three sources, the arteries, veins, or capillaries, and its source can usually be determined by the color and manner in which the blood flows.

ARTERIAL: Hemorrhage from an artery is very dangerous, and life is often quickly lost; here it is that knowledge, judgment, and coolness is called for. The blood from an artery is bright red and flows in jets, each jet or spurt corresponding to the beating of the heart.

VENOUS or CAPILLARY HEMORRHAGE is marked by the slow steady flow of blood, it being darker than the arterial blood.

Treatment.—The first thing to do in all cases of hemorrhage is to elevate the bleeding part, whenever possible, above the level of the heart. If in the hand or arm, raise above the head. In this position the blood has to travel up hill, and therefore reaches the wound with less force, which means slower escape of blood, and a greater likelihood of forming a clot at the mouth of the wound. Cold water or ice applied to the parts assists, especially in wounds of small arteries, by contracting the vessels. Styptics should be applied to contract the bleeding surface* and to aid in forming a clot. Probably the best remedies for this purpose are tannic acid, subsulphate of iron, or a mixture of vinegar and water, about one part of the former to four of the latter. Pressure on the artery at the wounded point or above the wound is a very simple expedient and will, in nearly all cases, prove of very great value in controlling the flow of blood until more permanent means can be employed. Several methods follow.

Pressure with the fingers is made by placing one finger just

---

* See page 117.

above and one below the wound, and crowding the edges together firmly.

Pressure by compress and bandage consists of making a compress of some soft material, such as muslin, linen, cheese cloth, etc., folded into suitable shape and bound securely upon the wound with a bandage.

Pressure above the wound may be made by the finger, or when the wound is in a limb a tourniquet may be employed, which is readily made by tying a handkerchief around the limb loosely, with secure knots, then pass a pencil or knife through the handkerchief and twist in such a manner as to tighten it sufficiently to stop the bleeding. This may be continued until the surgeon has ligated the artery, which, of course, should be as soon as possible.

HEMORRHAGE FROM THE NOSE is not infrequent and is usually not dangerous. It is generally due to the rupture of some of the capillaries of the lining membrane of the nose. But there are times when the bleeding is due to the rupture of a small artery, possibly from ulceration, when it is so profuse as to threaten life. In such cases, which are very rare, medical aid should be summoned at once.

Treatment.—Have the patient to sit upright, and to hold the hands above the head, at least the one on the bleeding side. Take a towel and wring out of ice water or place in it finely cracked ice, and wrap it around the neck. If bleeding continues, have patient snuff up from the hand or inject into the nostrils, with a small syringe, ice water and alum, a teaspoonful of alum to a half glass of water. Vinegar, too, used in the same manner, will usually stop the flow of blood. But if these means should fail Dr. Butler's plan will be found very good. He says: "Take a piece of cotton wool as large as the first joint of the thumb, tie a thread around its middle, soak it in alum water, or, if that is not at hand, oil it with sweet oil or vaseline and plug the nostril. This is best done by pushing the cotton with a screwing or twisting motion, until firmly lodged. The thread serves to draw it out when required."

HEMORRHAGE FROM THE GUMS can usually be controlled by syringing the mouth or sockets (if teeth have been extracted) with warm water to remove clots, then rinse the mouth freely with ice water and alum. If this should fail some good styptic should be applied. See page 63, "Hemorrhage after Extraction."

HEMORRHAGE FROM THE STOMACH may be caused by inflammation, cancers, corrosive poisons, etc., or may occur without any apparent cause. It is, however, not a very frequent occurrence. The blood in such cases, when vomited, is usually of a dark brown color, resembling coffee grounds, unless it has very recently and suddenly escaped into the stomach, when it is of a bright red color. Bleeding from the stomach should not be mistaken for bleeding from the lungs. It is therefore important to remember these facts, that blood from the stomach is usually of a dark brown color and is vomited and not frothy, while blood from the lungs is of a *bright red color and is frothy*, also *that it is coughed up, not vomited.*

Treatment.—Rest in bed should be insisted upon. Keep the patient calm, as excitement increases the heart's action and the amount of blood; for the same reason, stimulants should not be administered. The patient should be given small pieces of ice to swallow, and teaspoonful doses of vinegar may be given every ten minutes. Also ice-cold cloths may be placed over the stomach.

HEMORRHAGE FROM THE LUNGS is usually caused by consumption and is rarely if ever fatal, except sometimes in the last stages of the disease. It is, therefore, not necessary to apprehend immediate death, as is so often done.

Treatment.—The best treatment to pursue, until a physician arrives, is to place the patient in a reclining position, but not with the head low, give small lumps of ice to be swallowed, and let him eat a teaspoonful of salt with the same quantity of vinegar. Salt absorbs water from the blood, and thus tends to relieve hemorrhage by thickening the blood. Also, if patient is not too weak, cloths wrung out of ice water may be applied to the chest and neck.

Intoxication is usually caused by an excessive use of alcoholic beverages. It sometimes resembles apoplexy, and great care should be taken in determining the disease, as this mistake has been made with very embarrassing results. It should be remembered that in drunkenness there is a helplessness on both sides alike, but no paralysis, that there is usually some sensation displayed by touching the eyeball, and that the patient can be aroused from the stupor; also, that the odor of liquor can be detected upon the breath, though this might be the case in apoplexy.

Treatment.—A teaspoonful of aromatic spirits of ammonia in a half or two-thirds of a glass of water is a useful corrective and stimulant. If this is not at hand, a large draught of vinegar often does much toward sobering an intoxicated person. In an extreme case, where respiration has ceased, or where there is evidence of collapse, artificial respiration should be resorted to, heat should be applied to the body, and copious draughts of hot coffee should be administered.

Nausea, if caused by slight indigestion, can usually be corrected by taking a teaspoonful of baking soda and the juice of one lemon in a half of a glass of water; stir and drink while foaming. Or take a teaspoonful of aromatic spirits of ammonia in a third of a glass of water. If the nausea is due to something objectionable to the stomach, the soda water will usually give relief by causing vomiting.

Poisons.—The old maxim, "an ounce of prevention is better than a pound of cure," is surely applicable to the subject of poisons. In the first place, all dangerous articles should be kept out of the way of children; then, all bottles, etc., containing anything of a toxic nature should be distinctly marked, "poison." A very good plan is to have such bottles marked by a ball and chain, which may be procured at the drug store; this will give warning in the dark as well as in the light. When it is discovered that a poison has been swallowed, some one should be dispatched for a physician at once; meanwhile, treatment must be directed toward getting rid of the poison before it takes effect.

First administer a prompt emetic. Some poisons, by their irritating effect, naturally produce vomiting, so that with a little encouragement the stomach will be thoroughly evacuated. Where this is not the case the emetic will provoke expulsion of the matter. A very good emetic for such an emergency is lukewarm water in quantity, or a tablespoonful of mustard or salt to a pint of warm water. As this is no occasion for fastidiousness, any water that is at hand may be used; if soapy, and the hands have been washed in it, use it, as by its very repulsiveness it may act more quickly than anything else; the patient should be urged to drink freely until he can contain no more, and be made to vomit over and over again.

This sometimes leaves the patient much depressed in body and mind, showing signs of collapse. In such a case, some mild stimulant may be given; hot tea is probably one of the best, as it is also a chemical antidote to many poisons. Strong, hot coffee is also good. To either of these a teaspoonful of brandy may be added. The patient will, of course, be in bed, and it should not be forgotten that warm coverings are indispensable. Hot bricks and hot-water bags, or bottles, may be brought into requisition. Again, where it is known that poison has been taken, and especially if it is one of the more active and corrosive, an antidote to counteract the action of the poison should be administered before or after the emetic. The following list of the more common poisons, with their antidotes, etc., will be useful for ready reference :—

SPECIAL POISONS AND READY ANTIDOTES.

POISON.	TREATMENT.
Acids—Sulphuric, Nitric, Acetic, Oxalic, Muriatic or Hydrochloric, ...	Give an alkali; such as powdered chalk, plaster, lime-water, as much as patient can swallow; or lime scraped from the plaster or white-washed wall, stirred in a cup of water. A tablespoonful of strong soapsuds, etc.
Alkalies—Potash, Lye, Soft Soap, Strong Ammonia or Hartshorn, ...	Give an acid—vinegar, lemon-juice, sour cider, etc. Acids and alkalies neutralize each other—that is, combine to form harmless *salts*.
Arsenic, ...	Milk and raw eggs, or flour and water, or lime-water and oil, and after patient has vomited freely, follow with a dose of castor oil.
Carbolic Acid, ...	There is no chemical antidote, but the stomach should be protected and vomiting encouraged by giving mucilaginous drinks, flour and water, and oil freely—olive, linseed, or castor.
Chloral, ...	Treatment same as opium.
Chloroform, Ether, etc., . . .	Loosen clothing; sprinkle cold water on the face; suspend the patient by the legs; artificial respiration, as for drowning.
Iodine, ...	Starch and water; boiled or baked potatoes.
Lead—Sugar of Lead, ...	Epsom salts; after vomiting freely give dose of oil.
Mercury—Bichloride of Mercury or "Corrosive Sublimate,"	Albumen, uncooked white of eggs, wheat flour, milk, etc.
Opium—Morphine, Laudanum, Paregoric, ...	Induce vomiting first. There is no chemical antidote, but strong coffee, pain, *motion* counteract its effects. In extreme cases, in addition to above treatment, cold water should be dashed on the face and chest, and artificial respiration and the battery resorted to.
Phosphorus, ...	Provoke vomiting. Teaspoonful doses of turpentine, mixed with magnesia; but *no* oil, it favors the action of phosphorus.
Prussic Acid, ...	Induce vomiting first, and give teaspoonful of ammonia in water.
Silver—Nitrate of Silver, "Lunar Caustic,"	Large teaspoonful of salt in a glass of water; vomiting.
Strychnine, ...	Induce vomiting first; give a purgative; secure absolute quiet, in a dark room.

There are other poisons, such as Alcohol, Aconite, Belladonna, etc., which need not be classified here, as vomiting thoroughly, followed by a mild stimulant and rest, is all that is needed.

Poisonous Bites.—By this we mean bites of rabid or venomous animals and the stings of insects.

Snake Bites.—Tie a cord, or a handkerchief twisted into a cord, tightly around the part just above the wound. Enlarge the wound by making a cross cut through the centre of the bite with a pen-knife. This will encourage bleeding, and will expose the wound more thoroughly for the later steps in treatment, which are as follows: Draw the poison from the wound by means of suction with the mouth, unless the mouth be sore, or by taking a wide-mouthed bottle, and after saturating a piece of cotton or paper with alcohol or benzine, set it on fire and dip it into the bottle. As soon as the flames begin to die out, quickly invert the bottle over the wound and press tightly against the skin to prevent the admission of air. This will extract the venom and blood from the exposed vessels. Then heat a knitting-needle, piece of wire, or small blade of a knife to a white heat, and thrust it into the wound. At the same time large doses of whisky or brandy should be given and the patient kept under the influence of the stimulant until medical aid can be secured. Some, however, prefer to use aromatic spirits of ammonia instead of whisky, one drachm to a wineglassful of water.

Mad Dog Bites.—There are some physicians who claim that there is no such disease as hydrophobia; one author puts it in this way: "So-called hydrophobia exists exactly in proportion to the common belief in it," that is to say, the trouble is altogether mental. There is no doubt but that a great many deaths have been caused by fright and anxiety, but that all cases are spurious I am not prepared to believe, and think that prompt and heroic treatment should always be given.

As dogs are, with many people, daily companions, it is important to know the various symptoms of madness. The following is a résumé of the instructions issued by the Council of Hygiene, of Bordeaux.

SIGNS OF MADNESS IN DOGS.—" 1. A short time after the disease has been contracted, the dog becomes agitated and restless, and turns continually in his kennel. If unchained, he roams about aimlessly; he seems to be seeking something; then stands motionless, as if waiting; he starts, snaps at the air, as if catching a fly, and dashes himself, barking and howling, against the wall. The voice of his master recalls him and he obeys, but slowly, with hesitation and seeming regret.

"2. He does not try to bite, is gentle, even affectionate, and eats and drinks; but he gnaws his litter, the ends of the curtains, the padding of cushions, bed-coverlids, carpets, and anything which happens to be in his reach.

"3. From the movement of his paws along the sides of his open mouth, one might suppose him trying to free his throat of a bone.

"4. His voice is changed so markedly that it is impossible to overlook it.

"5. He becomes surly, and begins to fight with other dogs."

The symptoms, however, vary in different cases, and a change in the habits or manner of a pet dog should always be looked on with suspicion, and the animal should be chained for a while.

The probability of hydrophobia being communicated to persons bitten by a mad dog varies with the location of the bite. If it be in a part unprotected by clothing, inoculation is almost certain; in other parts, the chances depend on the thickness of the clothing, which wipes the virus from the teeth.

Treatment.—The treatment in case of mad dog bite must be altogether preventive, as after the specific symptoms manifest themselves the only thing to do is to keep the patient quiet by the administration of hypnotics, until death ensues. But this, of course, belongs to the physician. When one is bitten by a rabid dog, the same course should be pursued as directed for snake bite, *excepting the stimulants.* Then start at once for a hospital, where the patient can receive Pasteur's treatment by inoculation.

Bites and Stings of Insects.—Despite the current belief that the bite of the tarantula, centipede, and other insects are dangerous to life, experience proves that they are in nearly all cases comparatively harmless, causing only temporary pain and annoyance.

Treatment.—They may be treated with cold, wet applications; if nothing better is at hand, wet earth is good. The application of a few drops of hartshorn at the point where the sting entered will also give relief. It sometimes happens that a wasp or bee is swallowed in taking a drink of water hurriedly or in the dark. In such cases the fauces swell rapidly from the moment the sting enters the throat, which places the patient in danger of suffocation. This should be treated by the free use of a gargle of hot water and salt, pending the physician's arrival.

Sea-Sickness.—Sickness occasioned by the motion of a vessel at sea is often most distressing. The most efficient preventive or treatment is to take a seat near the centre of the vessel (if inclined to keep up) and, as the ship descends, take a full breath; wear a wide, firm belt around the stomach, eat lightly, plainly, and often; if the stomach is much disturbed, take the juice of one lemon in half a glass of water, with one teaspoonful of baking soda; stir and drink while foaming.

Sprains.—A sprain is a sudden overstretching and tearing of the ligaments which enter into the formation of a joint, as well as the tendons and muscles about the joint.

Treatment.—The joint should be soaked in water as hot as the patient can bear for twenty minutes or more, then rub gently with cosmoline, and apply a snugly fitting but not tight flannel bandage, and give the part as near perfect rest as possible.

Starvation.—When a person is found exhausted from starvation, he should be placed in a comfortable position and given stimulating, fluid food. Warm milk, soup, and hot coffee are among the best, given a little at a time, but often.

Strangulation.—In cases of strangulation, that is, the compression of the windpipe from the outside, as in hanging, etc., the treatment to be pursued is to remove the pressure at once, and re-establish the respiration as in drowning.

Suffocation.—Suffocation from foul air, noxious gases, etc., is caused by a poisonous gas known as carbonic acid. When it is desirable to enter a cellar, well, mine, etc., where there is a suspicion of foul air, a thorough examination should be made. Man cannot live in an atmosphere where a candle will not burn, animal life and flame being both supported by oxygen. The best test, therefore, is to lower an unprotected light where foul air is suspected; if the flame flickers and goes out, by no means enter.

Carbonic acid gas, being heavier than air, can readily be removed by the use of a pump, but, if this is not at hand, quick-lime (lime freshly burned), scattered about in large quantities, will accomplish the purpose.

Treatment.—In case of asphyxiation from noxious vapors the patient should be removed as soon as possible to fresh air, and natural respiration reëstablished, as directed for drowning.

Sun Stroke does not necessarily arise from undue exposure to the direct rays of the sun, but may proceed from a prolonged elevation of the bodily temperature, or from excessive heat encountered when the vital forces are near the point of exhaustion. It is generally preceded for some time, usually from one to three days, by pain in the head, a feeling of weakness, disturbance of the sight, and nausea. This attack, however, culminates, usually after the third day, in a loss of consciousness. The skin is intensely hot and dry, the temperature rising as high as 112. In fevers, if the temperature rises to 105 or 106, it is considered a severe case. It is, therefore, apparent that the patient is suffering from an excess of heat in the body. The thing to be done, then, is to lower the temperature as soon as possible. Every minute being valuable, the following treatment should be pursued:—

Treatment.—Send some one for a physician. Remove as much of the clothing as possible, and place the patient in a cool and airy place, indoors or out. Cold must then be applied to the head and body, not dashed or sprinkled, as that would only cause a needless shock; but towels wrung out of ice-water, and frequently renewed, should be placed upon the head, cracked ice placed in the arm-pits, and the body may be wrapped in cold, wet blankets. Continue this treatment until the physician arrives or the patient shows signs of consciousness, then discontinue, unless consciousness should again be lost or the surface of the body becomes very hot. *Never in such cases administer a stimulant.*

Wounds.—In surgery, wounds are divided into three classes, according to their cause, namely, *incised*, *lacerated*, and *contused*. There is a subdivision of this classification, of course, but this is all that is necessary for our present purpose.

Incised Wounds are those made by sharp-cutting instruments, making what are called clean cuts; that is, there is no tearing or bruising, but the edges are clean cut and the surface smooth.

Treatment.—If the wound is simple and small, the only treatment that is required is to cleanse the edges and apply adhesive plaster and perhaps a bandage. But where the wound is more extensive and serious, the edges should be brought firmly together, if possible, and held in that position by adhesive plaster and bandages. But when this will not answer, hold the parts together with the hands until the physician arrives. Dr. Dulles says: "In case an entire part be cut off, as an ear, or a nose, or a toe, or a finger, it should be cleansed with lukewarm water and put in its place, leaving to the surgeon the decision whether or not it would be worth while to try to save it. Some very remarkable cases of reunion of such parts are on record, and an attempt to save them is not to be lightly rejected."

Lacerated Wounds are made by blunt tearing instruments, such as dull tools, pieces of machinery, nails, hooks, etc. These wounds are rough and ragged and usually bleed but little. They should be given surgical skill, but until this can be secured, the torn parts should be cleansed by a stream of lukewarm water, then brought as near their natural position as possible and covered with a cloth soaked in phenol sodique, tincture of marigold, or laudanum, and wrapped lightly. If no good remedy is at hand, wrap in cloth wrung out of cold water. If the patient seems much depressed, administer a little brandy or wine.

Contusions are what are commonly known as bruises; they are usually caused by some blunt instrument or a fall; the skin is not torn through, but is often discolored, which is due to the rupture of the capillaries, allowing the blood to escape into the surrounding tissues—the familiar black and blue appearance of a bruise.

Treatment.—Such wounds are best treated by directing upon the wounded part a stream of water, as hot as the patient can bear it, for several minutes. This will favor the carrying off of the escaped blood. Then, after bathing the part freely with phenol sodique or laudanum, wrap in hot, wet cloths.

WEIGHTS AND MEASURES.

APOTHECARIES' WEIGHT.

20 grains (gr.)	make 1 scruple, ...	sc. or ℈
3 scruples	make 1 drachm, ...	dr. or ʒ
8 drachms	make 1 ounce, ...	oz. or ℥
12 ounces	make 1 pound, ...	lb. or ℔

SCALE OF COMPARISON.

℔.		oz.		dr.		sc.		gr.
1	=	12	=	96	=	288	=	5760
		1	=	8	=	24	=	480
				1	=	3	=	60
						1	=	20

APOTHECARIES' OR WINE MEASURE.

60 minims (♏)	make 1 fl. drachm, ...	fl. dr. or fʒ
8 fl. drachms	make 1 fl. ounce, ...	fl. oz. or f℥
16 fl. ounces	make 1 pint, ...	O.
8 pints	make 1 gallon, ...	C.

SCALE OF COMPARISON.

Gallon.		Pints.		Fl. ounces.		Fl. drachms.		Minims.
C.		O.		f℥.		fʒ.		♏.
1	=	8	=	128	=	1024	=	61440
		1	=	16	=	128	=	7680
				1	=	8	=	480
						1	=	60

TROY WEIGHT.

24 grains (gr.)	make 1 pennyweight, ...	dwt.
20 pennyweights	make 1 ounce, ...	oz.
12 ounces	make 1 pound, ...	℔.
3½ grains	make 1 carat (diamond weight),	. k.

APPENDIX.

SCALE OF COMPARISON.

lb.	oz.	dwt.	gr.
1	12	240	5760
	1	20	480
		1	24
		1 k.	3½

AVOIRDUPOIS WEIGHT.

16 drachms (dr.) make 1 ounce, ... oz.
16 ounces make 1 pound, ... lb.
25 pounds make 1 quarter, ... qr.
4 quarters make 1 hundredweight, ... cwt.
20 hundredweight make 1 ton, ... T.
100 pounds make 1 cental, ... C.

SCALE OF COMPARISON.

T.	cwt.	qr.	lb.	oz.	dr.
1	20	80	2000	32000	512000
	1	4	100	4000	25600
		1	25	400	6400
			1	16	256
				1	16

A gallon contains **eight pints.**
A pint contains sixteen **fluid ounces.**
A fluid ounce contains **eight fluid** drachms.
A fluid drachm contains sixty **minims** (♏).

APPROXIMATE MEASUREMENT.

A wineglass **contains two fluid ounces.**
A teacup **contains four fluid ounces.**
A teaspoon of powder contains one-half drachm.
A tablespoon of powder contains two drachms.
One drop of essential oils contains one-half minim.
One drop of water contains one minim.

APPENDIX.

From Gould's New Medical Dictionary.

The following table of approximate and exact equivalents of the metric and common weights and measures may prove serviceable.

LENGTH.

Unit of Measurement.	Approximate Equivalent.	Accurate Equivalent.
1 inch,	2½ cubic centimeters,	2.539
1 centimeter ($\frac{1}{100}$ meter),	0.4 inch,	0.393
1 yard,	1 meter,	0.914
1 meter (39.37 inches),	1 yard,	1.093
1 foot,	30 centimeters,	30.479
1 kilometer (1000 meters),	⅝ mile,	0.621
1 mile,	1½ kilometer,	1.609

SURFACE.

Unit of Measurement.	Approximate Equivalent.	Accurate Equivalent.
1 hectare (10,000 sq. meters),	2½ acres,	2.471
1 acre,	⅖ hectare,	0.404

WEIGHT.

Unit of Measurement.	Approximate Equivalent.	Accurate Equivalent.
1 gramme,	15½ grains,	15.432
1 grain,	0.064 gramme,	0.064
1 kilogramme (1000 grammes),	2⅕ lbs. avoirdupois,	2.204
1 pound avoirdupois,	½ kilogramme,	0.453
1 ounce avoirdupois (437½ grains),	28⅓ grammes,	28.349
1 ounce, Troy or apothecary (480 grains),	31 grammes,	31.103

BULK.

Unit of Measurement.	Approximate Equivalent.	Accurate Equivalent.
1 cubic centimeter,	0.06 cubic inch,	0.061
1 cubic inch,	16⅓ cubic centimeters,	16.386
1 liter (1000 cubic centimeters),	1 U. S. Standard quart,	0.946
1 United States quart,	1 liter,	1.057
1 fluid ounce,	29½ cubic centimeters,	29.570

INDEX.

A.

Abbreviations, 8.
Abrasion of the teeth, 33.
Abscesses, 28.
    alveolar, 28,
    of the antrum of Highmore, 57.
Absorption of roots, 21.
Acids, their effect upon the teeth, 65,
    antidotes for, 149.
    burns from, 132.
Aconite, 73.
Æther, 74.
Alcohol, 88.
Alkalies, 149.
    antidotes for, 149.
Alum, 116.
Amelloblasts, 9.
Ammonia, 90.
Anæsthesia, treatment of dangerous symptoms, 77.
Anæsthetics, general, 74.
    local, 74, 84.
Analgesics, 71.
Anatomy of the teeth, 16.
Angle's method of fixation, 45.
Anodynes, 71.
Anomalies of the teeth, 24.
Antidotes, 148, 149.
Antipyretics, 110.
Antipyrine, 110.
Antiseptic mouth washes, 40.
Antiseptics, 121.
Antizymotics, 121.
Antrum of Highmore, diseases of, 57.
Apoplexy, 130.
Aristol, 127.

Arsenic, 102.
Articulation of the teeth, 19.
Artificial palates, 60.
    respiration, 135.
Astringents, 114.
Atropine, 72.

B.

Bandages, 141.
Belladonna, 72.
Bichloride of mercury, 122.
Bites, dog, 150.
    serpent and insect, 150, 152.
Bleeding, 143.
Blood stasis, 28.
Bromide of ethyl, 83.
Bromine, bromides, 69.
Bruises, 155.
Burns, 130, 131.

C.

Calcareous deposits, 63.
Calcifications of the teeth, 11–13.
Camphor, 91.
Capsicum, 94.
Carbolic acid, 120.
Caries of the teeth, 37–39.
    in relation to sex, 38.
Catalepsy, 132.
Cathartics, 129.
Caustics, 118.
Cementoblasts, 12.
Cementum, 12, 14.
Chemical irritants, 111.

Choking, 132.
Chronic inflammation, 26.
Chloral, 70.
Chloride of ethyl, 86.
Chloroform, 78.
  therapeutic uses of, 79.
Chloride of zinc, 106.
Cimicifuga, 99.
Cinchona, 96.
Citrate of magnesia, 129.
Cleaning teeth, 164.
Cleft palate, 58.
Cocaine, 84.
Congestion, 28.
Cold and heat, 110.
Contusions, 155.
Convulsions, 132.
  in teething, 20.
Corn ergot, 118.
Corrosive sublimate, 122.
Crowns of teeth, 18.
Cuts, 154.
Cystic tumors, 27.

D.

Deciduous teeth, germination of, 9, 10.
  decalcification of, 21-23.
Dental caries, 37-39.
  relative location, 38.
  medicine, 68.
  pathology, 25.
  periostitis, 34.
  pulp, 14.
  therapeutics, 25.
Dentinal fibrils, 12.
Dentine, 14.
  calcification of, 11.
  formations, 33.
  organ, 9.
Dentition, 19.
  lesions incident to, 20.
Development of teeth, 9.
Digitalis, 98.
Disease, 25.

Diseases of dental pulp and membrane, 31.
  of hard dental structure, 37.
Disinfectants, 121.
Dislocation of the inferior maxillary, 55.
  treatment of, 56.
Dislocations, 133.
Dover's powders, 71.
Drowning, treatment of, 133.

E.

Emergencies, 130.
Enamel, 9, 12.
  formation of, 9, 11.
  organ, 9.
Enamelblasts, 9.
Epilepsy, treatment of, 137.
Epithelium, 9.
Epsom salts, 129.
Epulis tumors, 27.
Ergot, 118.
Eruption of the teeth, 19, 23.
Escharotics, 118.
Ether, 74.
  administration of, 75.
  action of, 76.
Ethyl chloride, 86.
Etiology of dental caries, 37.
Eucalyptus, 100.
Eugenol, 95.
Exhaustion from heat, 140.
Exostosis, 35.
Extraction of teeth, 61.
  hemorrhage following, 63.
  indications justifying, 61, 62.

F.

Fainting, 140.
Ferrum, 101.
Fistula, 28.
Follicles of the teeth, 9, 11.
Foreign bodies in the throat, 132.

Formula of permanent teeth, 17.
  of temporary teeth, 16.
Fowler's solution, 141.
Fractures of alveolar process, 40.
  of maxillæ, 40.
  treatment of, 41–53.
Freezing, 143.
  the gum before extracting, 87.

G.

Gallic acid, 116.
Gangrene of the pulp, 34.
Gas, nitrous oxide, 80.
  administration of, 82.
  liquefied, 81.
  mortality of, 83.
Gases, noxious, 153.
Germination of the teeth, 9.
Glacial phosphoric acid, 109.

H.

Hæmostatics, 117.
Heat-stroke, 140.
Hemorrhage, 143.
  after extraction, 63.
  from lungs, 146.
  from nose, 145.
  from stomach, 146.
Hydrochloric acid, 109.
Hydrogen peroxide, 122.
Hydrophobia, 150.
  treatment of, 151.
Hypercementosis, 35.
Hypertrophy, 26.
  of the pulp, 33.
Hypnotics, 68.

I.

Induration, 27.
Inflammation, 25.
  acute, 25.
  chronic, 26.

Inflammation of the pulp, 31.
  of the pericemental membrane, 34, 35.
  of the temporo-maxillary articulation, 56.
Injuries and diseases of maxillary bones, 40.
Interdental splints, 43.
Interglobular spaces, 14.
Intoxication, 147.
Iodine, 112.
Iodoform, 126.
Iron, 101.
Irritants, 111.

L.

Lancing of gums, 21.
Laxatives, 129.
Lime water, 40.
Lunar caustic, 119.

M.

Magnesia, 129.
Materials used for splints, 141–143.
Mechanical irritants, 112.
Monsel's solution, 102.
Morphine, 71.
Mouth washes, 40.
Mucous deposits, 66.
  effects upon the teeth, 66.
Myrrh, 93.

N.

Narcotics, 68.
Nausea, treatment of, 147.
Necrosis of the jaws, 54.
Nervous irritants, 112.
Nitrate of silver, 119.
Nitric acid, 108.
Nitrite of amyl, 92.
Nitrous oxide, manner of preparation, 80.

INDEX.

Nitrous oxide, as an anæsthetic, 81.
    mode of administering, 82.
    mortality of, 83.
Nodular dentine, 33.
Nut-galls, 115, 116.
Nux vomica, 97.

O.

Obturators, 61.
Odontoblasts, 12.
Odontalgia, 33.
Oil of cloves, 94.
Oil of eucalyptus, 100.
Opium, 71.
Oxychloride of zinc, 106.
Oxyphosphate, 106.

P.

Pain after filling, 32.
Papilla, dentinal, 9.
Paralysis, 130.
Peppermint, 95.
Peridental membrane, 16.
Pericementitis, 34.
Permanganate of potassa, 125.
Permanent tooth follicles, 11.
Peroxide of hydrogen, 122.
    tests of, 124.
Persulphate of iron, 102.
Phenol sodique, 121.
Phosphoric acid, 108.
Physical effects of anæsthesia, 76, 77, 83.
Physiology of the teeth, 12.
Pitch, 114.
Pitted teeth, 14.
Poisons and their antidotes, 148, 149.
Polypus, 33.
Prophylaxis, 39.
Proximal surfaces, 37.
Pulp, calcification of, 15.
    devitalization of, 32.

Pulp, diseases of, 31
    exposure of, 32.
    gangrene of, 34.
    hypertrophy of, 33.
    inflammation of, 31.
    irritation of, 15.
    nodules, 33.
    structure of, 14.
Purgatives, 129.
Pus and pus formation, 26.
Pyorrhea alveolaris, 31.

Q.

Quinine, 96.

R.

Removal of temporary teeth, 21.
Relation of calcareous deposits to disease, 66.
Resolution, 26.
Respiration, artificial, 135.
Resuscitation from drowning, 133.
Rhubarb, 129.

S.

Saliva, 64.
Salivary calculus, 63.
Sanguinary calculus, 65.
Scalds, treatment of, 131.
Sea-sickness, 152.
Second dentition, 23.
Sedatives, 110.
Senna, 129.
Setting of fractures, 41, 141.
Shedding of temporary teeth, 21
Snake bites, 150.
Splints, 45, 141.
Sprains, 152.
Staphylorraphy, 59.
Starvation, 153.
Stimulants, 88.
Strangulation, 153.

INDEX.

Structure of the teeth, 12.
Strychnine, 98.
Styptics, 117.
Suffocation, from gas, 153.
Sulphate of iron, 102.
    of magnesia, 129.
Sulphuric acid, 106.
    aromatic, 107.
Sun-stroke, treatment of, 153.
Supernumerary teeth, 24.
Suppuration, 26.
Surface of teeth, 18.

T.

Tannic acid, 115.
Tartar of salivary calculus, 63.
Teeth, articulation of, 19.
    calcification of, 11, 13.
    crowns of, 18.
    development of, 9.
    exposed pulp of, 33.
    extraction of, 61.
    follicles of, 11.
    germination of, 9.
    occlusion of, 19.
    permanent, eruption of, 23.
    primary, eruption of, 19.
    relative proportion of roots and crowns of, 16, 17.
    supernumerary, 24.
    surfaces of, 18.
    temporary, 16.
      absorption of, 21.
    classification of, 16, 17.
    eruption of, 19, 23.
Teething, 20.
    convulsions in, 20.
    deaths from, 20.

Teething, treatment during, 21.
Therapeutics of caries, 39.
Third dentition, 24.
Tonics, 96.
Tooth powder, 40.
    pulp, 14.
Tourniquet, improvised, 145.
Tumefaction, 27.
Turpentine, 114.

U.

Ulceration, 30.
Use of tooth brushes, 39.

V.

Vascular tumors, 28.
Vent holes for gases, 35.
Velum, artificial, 60.
Vital force, 20.
Vomiting, to produce, 148.
    to allay, 147.

W.

Weights and measures, 156.
Wounds, contused, 155.
    incised, 154.
    lacerated, 155.
    punctured, 155.

Z.

Zinc, 105.
    chloride, 106.
    oxy-chloride of, 106.
    oxy-phosphate of, 106.

www.ingramcontent.com/pod-product-compliance
Lightning Source LLC
Chambersburg PA
CBHW030254170426
43202CB00009B/735